# SCREENS AND THE EGO

A
MEDITATION
ON GEN Z

# JANE-MARIE AURET

DEFIANCE PRESS
& PUBLISHING

# SCREENS AND THE EGO

**DEFIANCE PRESS**
& PUBLISHING

ISBN-13: 978-1-955937-77-1 (Paperback)
ISBN-13: 978-1-955937-76-4 (eBook)
ISBN-13: 978-1-959677-62-8 (Hard Cover)

Edited by R.D. Black

Workshopped by R.D. Black, Rummana Habib, Ph.D. pending, Maya Foster, Ph.D. pending, Pastor Jim Mong, and Mark Walker

Informally reviewed by many other members of my writing community.

Pre-publication libel review completed by the Cronin Law firm.

Cover art by Paul Vermeesch

Website by Cherry Ng-Sutowski

The Cronin Law Firm performed a pre-publication libel review such that my writing would be compliant.

Published by Defiance Press & Publishing, LLC

Bulk orders of this book may be obtained by contacting Defiance Press & Publishing, LLC. www.defiancepress.com.

Public Relations Dept. – Defiance Press & Publishing, LLC
281-581-9300
pr@defiancepress.com

Defiance Press & Publishing, LLC
281-581-9300
info@defiancepress.com

# DEDICATION

*Dedicated to my grandma the poet*

# TABLE OF CONTENTS

My brother recites the *Iqama* in a Russian accent, inserting a ye- in the middle of *la allaye ilye Allah*. His voice is our father's. Baritone. The majority of the other students in the hall don't speak Arabic well, and even if they hear the bad pronunciation, they dismiss it because they know where we were born. But I know my brother knows the right vowels. I know that Ilsur purposely recites like our father.

In Uzbekistan in the 90's, our mom learned to load guns in high school. She told me about it as she was making soup, her second American baby on her hip. "They gave each student a Tokarev – even the girls – and they showed us."

A Tokarev is a rifle. There's a joke about guns that goes with that. What has a barrel, a trigger, but doesn't fire? A Soviet rifle. The joke is that Soviet machinery doesn't work.

What has three wheels and a cannon?

A Soviet tank that was built for four wheels.

I took the baby from her so she could ladle the soup into bowls. The American baby's name was Eliza Hubbard, my half-sister. She was Richard Hubbard's daughter. Our mom, Arslana, which means lioness, and our Dad, Ilsur IV, which means "the Hero of the Motherland," had three children in our former country. We are: Chechek, which means flower, because she is so delicate and pretty, Ilsur V, my brother, and Ilnara, me, which is the word 'flame,' with a feminine ending.

When Ilsur and I were seven, our father died. Our mother needed to leave the country, so she made a profile on a dating website. Richard Hubbard matched her. Richard was awkward and nearly 400 pounds. About four years after some woman whose name he will not say left him for another man, he made the profile, matched with our mom, and brought us here. But he treated us well enough. He brought us to Atlanta, Georgia when Ilsur and I were nine. Chechek was eleven. We touched down in the States and, first thing, my mom took Richard upstairs to get her pregnant. Turns out that she did not need to secure leverage like that. Richard truly had no ill intentions. Richard and our mother had two babies, Blake, who is ten, and Eliza, the baby. They both have white-blonde hair like their father.

Richard Hubbard was a Southern Baptist who works at Technology Services at Emory University, so all three of us, the children from the first marriage, went to college with a partial courtesy scholarship. Really, Chechek was the only one who deserved her acceptance.

But we're all at Emory, now.


"Why didn't you stay after prayer?"

"How much do I have to do for you before you're thankful?" I snapped, hopping onto my dorm bed and throwing a small pink blanket over my legs. We switch between Tatar, English, and Russian mid-sentence.

Ilsur bent his tall, hard body forward. "It's your duty to come." What he meant by this was, 'I told you to come, and I want you to treat me like I'm a man,' so instead of fighting him, I say, "You sounded just like dad."

This calms him down.

"Come here," I told my brother. "Turn around." On my half-lofted dorm bed, I could push on his shoulder blades. He hunched. When we were at home, our mom would constantly nag Ilsur, "Do you want to look like a coal miner on your wedding day? Walk down the aisle with a cane?" She never nagged like this to Chechek or me, just would smack us between the shoulder blades until we sat up. But she would never do that to Ilsur.

"I needed your support," he said.

"Look, I came to the prayer, which you know I don't like. And, anyway, the food they serve after service is disgusting and sugary."

Three weeks before Ilsur led prayer, he publicly screamed at the president of the Muslim Student Union, Haleema. Haleema had scheduled four Islam 101 seminars to explain the religion to non-believers over the course of the semester and spent more than one-half of the whole budget on food for those events. In the third session, Haleema said that Muslims are an ally to the queer community because, she said, "the Quran alKareem never mentions the word *homosexual*," which is true. Well, in the past fifteen years some exegetical scholars who write for western universities have argued this point, but those are peripheral, academic conversations. Either way, Islam gives explicit guidelines for marriage, and many, many people would disagree with Haleema's interpretation.

In the middle of Cox Hall, Ilsur got angry. His entire body oscillated. "Haleema, all you ever do is pander to American liberals!" He rubbed his right temple and his face flushed. "If you gave one-fifth of the energy that you spend seeking the approval of white American *jahil* to authentically practice your religion, then you *might* be fit to lead! But you don't! You misrepresent Islam so you can get popularity points? Well, you don't earn points in Heaven for explaining that you don't eat pork! You gave our religion out to people to pleasure themselves, so they can feel smug about being multicultural, but the beliefs of Islam do not bend for your politics! You *whored* our religion out to them. You gave it to them. What's wrong with you? You are not fit for your role!"

Haleema sobbed, "You called me a whore?!"

As Ilsur grew angrier, he rubbed his right temple harder. "No! I said you whored out our religion, because you gave it to them to pleasure themselves, which is much, much worse!"

Everyone curled up, cringing.

I care much more about my brother than Haleema, who is my gossipy lab partner in remedial biology for non-science majors. But I knew that I needed to intervene. Ilsur needed this club.

I went to lab the next day ready to make peace. "Gosh, I'm so sorry."

"What's wrong with him?!" Haleema huffed, slipping on her protective plastic lab goggles. "He's just not going to be allowed in the club anymore."

"You know, I understand why you would say that, but please give me a chance to explain." It was not in Haleema's leadership interests to foster a public schism with someone who had memorized the Quran. "You don't want people to have to choose sides." I pointed out. "He's pretty good at admitting when he has lost his temper. If I get him to apologize—*publicly* apologize—will you accept it?"

"I guess." Haleema begrudgingly agreed.

At the next meeting, Ilsur publicly apologized for his tone when he confronted Haleema. In return, Haleema apologized for misrepresenting the beliefs of all Muslims. She's an ally, "unlike some other people here." Haleema agreed to stop spending the budget on Islam 101 sessions. Ilsur would lead prayer.

And he keeps telling me that *he* needs *my* support.

After Ilsur's situation was solved, I went back to living my life, closing the door of my dorm room and pulling the shirt off Kyle Lancaster, admiring his muscle tone. I fell in love with Kyle last year. We first started talking in a class on Marx, Nietzsche, and Freud because Kyle backed me up when I asked, "What does he want instead of capitalism?" When the *Das Kapital* segment of the class ended, Kyle and I discovered Nietzsche together. One time we stood in the hallway for three hours just talking after class.

I have never been more authentically me than when I was naked on top of a man quoting Fredrich Nietzsche. We both loved *On The*

*Genealogy of Morals* like it was a person. "What do you think Nietzsche would say about the excessive number of pillows on my bed?"

"Well, he said that water animals must have felt a terrible heaviness when they came to land, so maybe your desire to swim in pillows is an attempt to go back."

"A primal urge?"

"An eternal return?"

I sat up like I was riding Kyle, and he stopped mid-sentence to stammer at the image of my chest. "I-I am sanctified in your presence," he whispered to himself, the good boy he is. "You are so classically beautiful. A master sculptor would create you in an attempt to make a perfect woman. Look at the ratio between this and this. I love that I can see your abs—my God!"

"Your God or your Ubermensch?" I teased, referring to The Great Man in Nietzsche.

"Both!" he said as he rubbed his finger in spirals around my belly button. I went back down and kissed under his ear, which made him twitch.

Lying on my bed, I thought about beds. Like, the concept of them. I thought of how many people in history slept on dirt with no blankets. In our house in Uzbekistan, properly middle class, we had thick mattresses which lay flat on the floor, maybe four inches off the ground. Our mom and dad slept in a bedroom behind the kitchen. Chechek and I shared a soft mattress in the second bedroom. Ilsur slept on a cot in the living room. To make his bed in the morning, Ilsur had to slip the mattress behind the back of the couch.

When we moved into Richard's house in Atlanta, Richard had bought plush, tall American beds for us. Ilsur took his mattress off the bed frame and laid it flat on his bedroom floor.

The next day I opened my eyes, Kyle sprawled out next to me, and I remembered how much I hated school. I texted Ilsur to see if he could meet up to study together. He texted back, "sure." I kissed Kyle while he still slept and set off for my brother.

I walked into Ilsur's apartment proposing, "Well, see, I think it would be beneficial for everybody to specialize their labor, as they say in economics!" Ilsur had earned the highest test score in the lecture hall of two hundred students for Intermediate Microeconomics. I handed him my homework.

"The school will stamp your record for two years if you cheat."

"See, this is why no one likes you," I explained, taking my binder back. "You're like Angela from *The Office* except six feet tall and bearded."

"Do you want something like that on your record for two years?"

"When will you remove the stick from your ass and decide to enjoy the life you have?"

Ilsur pursed his lips. "Fine, what do you need help with? Help isn't cheating."

I showed him the problem. He scoffed and started to write numbers down in rows. My brother is remarkably non-verbal.

"Here," he spat, showing me the completed problem and expecting me to understand his chicken scratch.

"Ahh, I almost understand. Could you, maybe, show me three more times?" I pointed to the remaining three problems.

"No. You can learn and do the work."

"But, I don't like learning. Nor do I like doing work. But I *do* like getting A's and being special, so, you see my predicament here."

"You're in a real pickle."

"UGH!" I grunted at him and walked out, my binder under my arm.

"Where are you going?"

I flicked my neck with two fingers, a hand gesture meaning 'to get hammered.'

We decided to throw a joint party for Ilsur's and my birthday at his place in the upper-class housing apartments on campus. Chechek and I planned to make a small dinner for us and our close friends, Damion, Claire, and Andrew.

Ilsur let me into the building and helped me carry the groceries up the open, concrete corridors. I came with black rye bread, sour milk from grass-fed cows, black tea, lamb, and vegetables. The simplest ingredients in Uzbekistan are only found in the most pretentious grocery stores in America. I began cutting the onions, carrots, and celery and searing them at the bottom of the soup pan in butter. The butter bubbled as it burned.

"Smells delicious," Ilsur complimented. He sat with his laptop in the living room, surfing 4chan's politics board, casually listening to a livestream from an Uzbek national living in northern Afghanistan. The speaker wanted his clip to circulate widely, so he needed to translate his message into many different languages. First, he made his case in heavily accented Russian, then Uzbek, then in Iraqi dialect Arabic, and then his friend would say what he had said in Dari, then in Pashto. Thus, it took a very long time for him to get through his speech, which reminded me of the slow, repetitive life in the pastoral, patriarchal, nearly pre-industrial culture I used to know, except for a smattering of Soviet factories in our cotton fields.

I heard the speaker of the livestream say in an accusatory voice: "They gave us children's toys designed to keep us fighting the Russians forever. Helping us was never their intention. They wanted our country to collapse on itself, and when it did, they invaded our country. We say no! They fight for a godless government. We say no! We fight for God alone, for there is no god but God!"

"Ilsur, turn that off," I ask in English.

"I'm listening to it," Ilsur huffed in Tatar, turning the volume down a little.

The YouTube livestream went on. "The *gharbi* soldiers don't know who is responsible for the twin tower attack. Their country is allied with the country which made the attack." The speaker meant Saudi Arabia. "They keep their soldiers ignorant. I sent my young son to ask a *gharbi* soldier what *Taliban* means. The soldier didn't know. Of course, he didn't know. Of all the things they teach their soldiers, they would never tell them that 'Taliban,' means 'Students.' Even *jahil* would feel apprehension to war with Students who were not responsible for the attacks…"

"Ilsur, please turn that off."

Ilsur defiantly ignored me.

Finally, I heard the livestream speaker conclude, "God is greatest!" in Arabic.

I hated that my brother consumed this media. It was so hard to pull him back from it. I yelled loudly from the kitchen. "You know that those people wouldn't accept you even if you went back, right? You are not one of them."

"I want to know what's going on in my father's country," Ilsur insisted, as he typed a post on 4chan.

"Ilsur, could you look at me when I talk to you, please? Please?"

"Ilnara! You do not get to tell me what information I am allowed to access! Do I tell you what you're allowed to watch? Half of what you watch is porn. Why do you think that you can forbid me from listening to a mere political podcast when I don't get to forbid you from watching actual trash?"

"Seriously? You're going to bring up *Game of Thrones*, right now?" I cut in English.

"You mean *Pornography With Swords*?" His real words were *Harram With Swords*, but I'm translating.

"Ilsur, I don't want to fight! Please don't let this become a fight. I just want to have a nice party with you! Please?" I started to cry.

"Goodness…alright," Ilsur closed his laptop and came to me in the kitchen. His face softened. "Ilnara? Hey, come here." I cried a little harder. "You're right, I don't need to listen to stuff that upsets you when you're around," my brother admitted.

"Thank you." I wiped a tear off the bridge of my nose.

Just as Ilsur and I were making up, Chechek knocked on the front door in time for the important ingredients to be added. I sighed, "Hi" as I let her in.

"Is the heat on too high?! Ilnara, you need to make sure it cooks slowly!"

"It's on medium-low."

"If the butter is burning, it's too hot! Can't you tell? Oof, you would have scalded the milk!"

I didn't fight her.

"Did you buy black tea?" We used black tea to flavor the soup broth.

"Of course."

"But not with any infusions? Last time you bought Earl Grey …"

"How about *you* make it?" I suggested, "Because *you're* just so *good* at this stuff." I handed the wooden spoon to her and pointed to the remaining ingredients on the kitchen counter. As Chechek cooked, I rolled small balls of paper and flung them at the back of her hair to see if they would stay.

The dinner was nice, except that Ilsur didn't say much and I loathed my control-compulsive, self-righteous, fake, obedient sister. I tried hard to ignore her backhanded comments so that I didn't embarrass myself in front of my friends. We called our mom on a group call, assuring her that we'd come back this weekend.

Then, I had a good time drinking with my friends back at my place, binge-watching the aforementioned *Pornography With Swords* to stick it to my brother.

When people hear our story, they're quick to judge Richard Hubbard, so I feel the need to defend him. Richard isn't intellectual, although he has a master's in computer science. Richard is not flexible. He is not okay with anyone rearranging his superhero collectables. But he's nice. I don't believe he has the capacity to be violent. Due to his weight, he couldn't win in a fight if he tried. But even still, when Richard throws dinner parties, he asks us to sit at the table and participate in the conversation. Richard conscientiously asks about our feelings and opinions. Chewing his vowels with the r-sound like a born-and-bred Georgia boy, Richard would ask, "What color would y'all like to paint your room?" or "Here's twenty dollars, go have a great time at the movies!" He paid for our ACT tutoring and our college. Richard loves his children Blake and Eliza very much. He even changed his baby's diapers to help our mom.

When our mom put herself on that dating website, she faced no insignificant risk of becoming a sex slave in the Balkans. She only agreed to be with Richard because he paid for the whole family's plane tickets to the United States at the same time. She had bet that even if Richard had lied about who he was, as long as the family was in the United States together, she could find a way to seek asylum. But she didn't have to do that. Richard secured our American citizenship quickly. He owned real estate in Atlanta with no mortgage—a small, old southern ranch he had inherited from his grandma, but that's good enough. Richard's job let us access an elite American college for less than one fourth of the regular price, and he voluntarily picked up the tab so that we could graduate debt free. Yes, Richard wanted our mom to cook, clean, and go to exercise classes because having a beautiful wife significantly elevated his social status, which to an American woman comes off as demeaning. But our mom had expected that arrangement. Honestly, given Richard's financial position, our mom would have suffered through nearly any indignity to keep their marriage alive. But Richard was not demeaning. Richard was supportive, faithful, and relatively empathetic for an awkward guy. I don't think of Richard as my father, but I love him for who he is.

Not Ilsur, though.

Ilsur could never.

It's difficult to explain the subtle adjustments which first generation immigrants like my mom never make; things deeper than learning that circling your upturned palm means 'get a move on,' in America. Richard wanted my mom to get a part time job so she wouldn't feel isolated. Richard advised, "Find something you think you'd like, Arslana." Our mom learned that our public elementary school was hiring classroom assistants. She decided that might be a nice fit.

The vice principal explained that she would be working with the kids who act out, so she was prepared to be patient and mature with them. However, in school in Uzbekistan there are strong negative social consequences for acting out. If you were that kid who spoke out of turn in class, you would not play tag with the cool kids. We empathized with our teacher, not the pupil. On the job, the school assigned our mom a boy named Josh who kept defying her by saying, "It's a free country! I don't have to put my pencil away. It's a free country!" Our mom gently redirected Josh back to his studies, but she could not for the life of her understand why all the other kids loved Josh.

A more egregious example is the day the teacher in her classroom accidentally sat in some green paint. The teacher didn't know about the giant stain on the seat of her pants. My mom felt reverent, horrified, and embarrassed for the teacher, but then she noticed that all the kids were giggling for some reason, and she kept looking around the room to see what could be so funny.

In that same vein of cluelessness, a couple years later our mom was unprepared to handle American teenagers. As Freud was never popularly disseminated in the East, people don't say things like, "Oh, I eat too much because my mom always gave me food to make up for it after my parents fought." People certainly don't say those things to excuse their habitually poor diet, and there's not as much pressure to define yourself, if that makes sense. So, naturally, as teenagers, Chechek and I launched a massive Freudian revolt against our mom. "Maybe I would be better at trusting people if you didn't always choose Chechek over me." "We wouldn't fight so much if you and Richard didn't fight so much." She and Richard didn't fight more than the average couple, but that's beside the point. Our mom did not know how to respond to all the blame.

Back in lab, I asked, "Hey, Haleema. I drew it. Can we turn this in?" I handed her a pathetic bar graph with Soil, Gravel, and Peanut Butter on one axis and Pea Plant Height in Centimeters on the other. Turns out peanut butter isn't the ideal medium to grow pea plants.

As Haleema checked over my work, I sent a picture of my yogurt-white breasts to Kyle in class.

Kyle wrote back, "You are simply the most magnificent woman in the whole world."

"How are you going to worship me tonight?" I texted.

"All the ways you want."

"OH MY GOD!" Haleema screeched. "What is that picture!?!"

"Nothing!"

"Are you sending nudes?!?"

"No!"

"Yes, you are!"

"No, I'm not!"

"Yes! You're sending nudes to Kyle Lancaster!"

"Please, say that louder! I'm pretty sure the people across the room didn't hear you, Haleema!" Immediately, this bitch asked, "How did this start? Why Kyle? Does Ilsur know?!?"

I put both hands on her shoulders. "Listen to me, listen. Ilsur cannot find out. It will hurt him very badly if he knows about this. Currently you, Kyle, and I are the only ones who know."

"Fine! Just tell me how this started!?!"

I rolled my eyes, wanting her to evaporate. But I said, "We met in philosophy class."

Kyle studied abroad in Morocco, spoke Arabic much better than I did, and asked me about religion like it entailed ingredients for a recipe, completely oblivious to the gravity of his question, or the gravity of my decisions with him. The truth was, I didn't believe that God spoke to a single Qureshi riding a camel in 610 AD. But, in defense of the story, they say the Angel Gabriel brought the Quran down from heaven, and it is impossible for man to write such beautiful prose without divine intercession. I agree with that.

As I lay next to Kyle, I suggested, "You should take a passage from Nietzsche and switch out every *Ubermensch* for Ilnara as you go down."

Kyle smiled and agreed. "Let me worship you." Then, he applied the same logic to what I said and began to recite, "*Bismi*-Ilnara *alRahman alRah*—" he substituted my name in a prayer meant for God.

I sprung up underneath him, "NO! Don't make that joke! That's so gross!"

"Ah! All right, sorry!"

"We don't make those jokes, Kyle!"

"I'm sorry, Ilnara."

I silently determined how to proceed. Kyle is half black, half white from Mississippi, and he often got away with racial jokes. I bent down and bit the meaty part just above his collarbone in half-joking retaliation.

"OWW!" Kyle squealed, wrenching his neck away. "Why are you so violent?!"

I laughed like it was a joke, but I felt so gross. He desecrated God with my name. In the back of my head, I imagined disintegrating into the dirt from which I was made.

I rolled over and said, "I'm too tired tonight," replaying in my head how Ilsur reacted to our mom.

When Ilsur was fourteen, he had screamed at our mother. "What kind of a woman SELLS herself on the internet!?!? A whore! A whore does that!" Every day I feel grateful that sentence came out in Tatar. If Richard had understood the words, he would have tried to come up on Ilsur, and Ilsur would have killed him.

Luckily, our mom just asked, "Alright, what should I have done? Should I have married in Uzbekistan or in America?"

"Don't act like this was a sacrifice for our family! You wanted to marry a rich American for your love of things and your general comfort."

"Good Lord, Ilsur, what do you want?! To go back? Because I'll buy your plane ticket right now! You can go sleep with mules while you try to find a job! Please! By all means! Rid us of your thunderstorm of a presence! This house will be much happier without you!"

In order to accurately recall our father, I have to admit that my memory of him shifted after he died. Prior to his death, I thought of our father as something between a brick wall and a gorilla. After he died, I saw the strongest man in the world floating up, up, up. It is as if my father lived in an ontologically different category of history and culture; time and space.

My dad slaughtered animals for their meat, and always had blood stains on his shoes. He cut down whole trees to chop into firewood. He could always point in the direction of Mecca. He insisted that women pray behind men. He drank scalding black tea. His arms were like cinder blocks. It would make sense to say that I stopped trusting the world when my father died, but actually, I stopped trusting the world *because* of him. I have one vivid memory that plays in my head on repeat.

In 2004, when we had just turned seven, about six months before he died, Dad took us downtown. The power grid failed frequently so everyone kept firewood and we, properly middle class, used lighter fluid to keep our oven working during the blackouts. We were so well off that we even had a home computer—a discarded monstrosity of a box with a tiny screen. No internet, obviously. But when the power was on, we could insert discs that showed pixelated images and interactive programs that taught us proper, sophisticated Russian. Paupers had to scrounge for fire kindling. I remember seeing little Asian-looking Uzbek kids picking up little twigs and thinking, *woah, they are so bum-backwards and poor,* blissfully unaware that's how most of the world thought of me.

Dad had taken us downtown because he believed that the shopkeeper had sold him watered-down lighter fluid. The shopkeeper functioned as a central trader of imported goods—and anything imported was better than homemade, so Dad needed to keep the relationship while also calling out the shopkeeper for tampering with his products. Dad finessed, "Hmm, my good friend, well it sure didn't light when I tried to light it. Who sold you this shipment? You may have been swindled."

"Surely not," the shopkeeper obviated from blame. "Perhaps you left it open outside?"

As they pretended that they weren't accusing each other, Ilsur and I walked the shelves of the shop and I noticed a little green handheld radio. Ilsur and I loved listening to stories over the radio. "No one would know if we took this," I suggested.

Ilsur's eyes bulged. "God would know!"

We had recently been taught the Revelation. There's a concept in Islam called *shafa'a*, intercession, which means that someone of high moral character vouches for you on Judgment Day. For example, if you have a family member who strays from God and then dies in a motorcycle accident without repenting, you can live a righteous life and then loan some of your good deeds on Judgment Day, because God is not indifferent to the pleas of those who honor Him with right conduct. Intercession will not work for those who do not pray, do not help the poor, live a life of indulgence, or deny the Day of Judgment. I didn't do any of that, so I curtly explained, "Dad will loan us some good deeds." After all, we wouldn't have to square this off until the Day of Judgment.

As Ilsur mulled over my foolproof exegesis, a uniformed police officer walked in. We gasped. We had watched rooms full of strong grown men shut up around uniforms before. Everyone knew you were supposed to stand still and quiet like a marbled polecat playing dead. In a panic, Ilsur whispered, "Quickly, quickly!" and I shoved the radio down my shirt. We watched if the officer saw, and the gigantic bulge on my abdomen gave it away. He yelled "Little thieves!" and approached us with his baton raised. Our father jumped in front of him like a bull. Boom.

"Respectably, sir," our father began, addressing the officer, but the officer swung his baton. Our dad dodged the blow and grabbed him by both his wrists to immobilize him without aggressing at him, which made the police officer look like a flayed lamb carcass on a rack. We felt proud. That dinky uniformed man was nothing compared to our big, strong father. Dad leaned in and whispered something to him. Whatever he said, reconciliatory or intimidating, it got the police officer to walk away. Then, dad looked back at me. I pulled the radio out of my shirt and meekly handed it to my dad. He gave it back to the shopkeeper. "They won't ever do that again," Dad apologized on our behalf.

We knew that we were in trouble on our way home. Ilsur pleaded, "It was Ilnara's idea! She was the one who stole it!"

"And you allowed it, boy? You can't even stop a little girl from doing something as stupid as stealing what she doesn't need?" Ilsur and I were born three hours apart, but Ilsur was born male.

When we arrived, Dad retrieved his slaughtering knives. He meticulously pulled the knives out of a long, thick leather pouch and laid them on the table: a tall one to sever the neck, a cleaver to hack the meat, and a hooked one to pull away the tendons of the kill. The knives were not what he needed. He picked up the empty leather pouch—about a foot long, heavy, and twice as thick as a belt. Chechek ran into our bedroom and shut the door. With his left hand, Dad held Ilsur by the collar of his shirt. With his right, he swung, and Ilsur's body swung. Ilsur looked like a towel blowing on a clothesline. Our mom screeched, "Enough! That's enough!" But, Dad ignored her. She didn't get to call the shots. When Dad decided Ilsur had had enough, he let him go, and Ilsur crumpled down.

Dad turned around and hit me with the empty leather knife pouch once. The heaviness was shocking. If my dad had hit any other part of my body, it could have injured me. He recognized that and said, "Your frame is too light, yet." Then, he dropped the knife pouch and pulled me down to finish with an open hand. Out of the corner of my eye, I saw a pile on the floor and realized it was my brother dry heaving. Never in my life had I been so grateful to be a girl.

Once I calmed down enough to process what had happened, I realized that I had humiliated Ilsur. I physically hurt him and I had made him look impotent to our father. That night, I went up to him on his bed in the living room and said, "You can take anything from my room you want," but I knew that I sounded pathetic.

"Never talk to me again." Ilsur gingerly rolled away from me. "I hate you."

"Look, I know this is my fault. Dad shouldn't have—"

"Who are you to say anything about what should or shouldn't happen, Ilnara? You don't know better than Dad."

"You can beat me up tomorrow, all right? Call it even?"

Ilsur silently considered my proposition. "Well, that would make us even." But then he decided, "No. I should have stopped you. I made the mistake. Go to bed, now, Ilnara."

I started to get up to do what Ilsur said, but our dad came out from the kitchen, popping a crouton in his mouth. "That's respectable, son. I heard your little exchange. That's good judgment. God the Highest respects a man who pays his debts himself."

Ilsur bashfully smiled, accepting the reconciliation.

"God the Highest is most merciful," Ilsur said back. I flared my nostrils. I had previously felt guilty, but now Ilsur just came off as a suck-up.

Dad turned to me, "I trust you won't make that mistake again."

Backing away slowly, I agreed, "No, I trust I won't."

My dad smirked at me for backing up. Amused, he asked, "Oh, you're angry at me, are you?" Hilarious. Really freaking funny. I pursed my lips.

My dad's face softened, realizing that I didn't understand why he would do something like that. "Ilnara, sweetie, do you know what would have happened if the *Cheka* got you?" *Cheka* is an old slang word for a division of the notorious Soviet secret police from the early 1940s. My dad meant it ironically, accusing the contemporary Uzbek police of being the corrupt backhand of Russia. "They would have thrown you in jail with no food. They wouldn't have cared that you're a child. There's no guarantee of a speedy trial. The cells are packed with people who've been starving out for weeks, and those pigs don't even separate men from women in their jails. Grown men would have done terrible, terrible things to you. Do you understand the danger you were in, girl?"

"Yes, I'm very sorry." No, I didn't understand at all. Surely, nothing in the world could have been more terrible than what happened three hours ago. It hadn't yet occurred to me that stealing is wrong. I still believed I was covered under intercession. "Very sorry," I repeated in a taciturn tone that indicated no remorse whatsoever.

My dad dragged me across the room as I screamed for mercy. He plopped me in front of the huge computer we had in the living room. Dad inserted a floppy disk and pressed buttons to start it. "A minute ago, I was glad you were indignant about your childish sores, but your insufferable attitude will get you killed." The computer screen turned on. He asked me, "Surely you've not heard news of our good brother in God Muzafar Avazov?"

I shook my head.

Dad pressed more buttons, and then an image appeared on the screen. It was of a naked man lying on a metal table. The lower half of his body seemed brown-red and blotchy. He had a thick belly but bone-thin, leathery legs. Dad said, "Muzafar wanted God in our politics, and for that, he was sentenced to twenty years at Jeslyk, the prison. Do you remember him? He came for tea."

I shrugged, not comprehending.

"At Jeslyk they boiled him from the waist down, Ilnara. You see the line around his legs? They suspended him from the ceiling and lowered him into a pot. The meat came off his thighs while he still lived."

I covered my eyes with my hands and tried to turn away. My father slapped my hands down, grabbed my ponytail, and shoved my face closer to the screen. "Do you understand what would have happened if they got you now, Ilnara?"

"Yes!" I shrieked. "I'm so sorry! Please!"

"Good." He pressed a button to make the image disappear. He opened his arms. Even though I hated him, I buried my face in him, bunching up the fabric of his shirt in my palms, sobbing. That image still haunts me today. It's real. If you Google "Muzafar Avazov," it comes up. Jeslyk functioned as a prison in northern Uzbekistan until 2019. Look it up.

My dad rocked me and tried to say, "It's okay. I've got you. I'll die a thousand deaths to make sure that never happens to us." Well, six months later, he died one death and never came back. Poof, gone.

I've spent many hours imagining what my dad would have been like as an American. Maybe he would have applied his compulsive black-and-white thinking to an accounting program instead of Islam. He probably would have joined a frat or the wrestling club. In a world with standard justice – like, if I had stolen the radio in America – my dad's only prerogative would have been to facilitate my moral development, and the story would have stopped at whatever he considered appropriate discipline … no TV for a week, whatever. It would not have occurred to an American to hold Ilsur responsible for my stealing. But Ilsur *was* responsible.

I want to be clear that I'm not explaining vapid cultural differences. I'm not telling you the distinction between spaghetti and tacos. I do not care what you think about the Tatar harmonic minor scale or big head jewelry or customs and rituals. 'Food and music' culture is a stupid person's understanding of culture. People who share the same economy, security of property, education system, and structure of labor, but who have different music and food, can be collapsed into a general theory of diversity and inclusion because none of their experiences actually create different senses of urgency towards good and evil. Two 30-year-olds living together in a safe city unmarried with no kids basically live the same life whether in Portland or Copenhagen, Muslim or Christian. There are many Muslims, like Haleema and her family, who have integrated enough that they pose little challenge to secular liberalism. But my father's culture cannot fit into the West's. It cannot integrate. You have to choose one or the other.

Want to see a real cultural difference? Ask a Masai person about animal rights. In that way, my brother and I were born into patriarchy as an ancient, pastoral social formation, integrated into pre-capitalist trade networks, spiced with the legacy of the autocratic Mongols. If groups of thieves were going around, we fought them off, not the police. The Soviets brought factories, put our women in those factories pretending it was for their equality, so they could be equally slaves as men. Stand and turn, stand and turn. We had to train all our wheat and cotton west of the Caucuses at no profit, theoretically in exchange for their measly handouts. Want to humiliate a man? Make him stand in a line begging to get just as big of a serving as the man in front of him. "You gave him more than me!" So, what did the men who still loved horses do? They guarded their women. They raised their sons in the image of men who love horses. Women are either protected by men or they are equal to men, but not both.

I say this to explain that our dad wasn't being mean to Ilsur or me. When you live in violence, men fight it. They just do. If I had been out with my mom the day I stole the radio, if my dad had been drunk, or even if my dad had just been inattentive, there's a very real chance that story would have ended with me raped in a jail cell before I knew about the third hole on my body.

The rapist wouldn't have been a pedophile in the American sense either. Equally evil, but not like the West thinks of that crime. He wouldn't have harbored an obsessive, deranged preference for children. He would have been a toothless, emaciated creature who knows his imminent death and deposits his genes into the closest approximation of a female around him. Ilsur needed to fear the Lord so that if he found himself in jail, he wouldn't succumb to his animal mind. Our dad thought that Ilsur would be a man of a violent land. If Ilsur didn't protect his family, then his wife and children would die. The cost is death. That is why Ilsur was responsible for my stealing.

You know how Americans read Thomas Hobbes's social contract in late high school? That's what it was. We submitted to the men in our lives on the condition they protect us. Dad was the government: our protection and therefore our authority. Feminism sounds so utterly stupid to my brother because men and women are not equal in combat. We're just not. A society which remembers combat as a frame of reference will not believe in equal gender roles. This is true inside American society as well as outside it. Rural, heartland Americans who are likely to serve in the military or police force don't accept feminism culturally, but rich, coastal Americans who are unlikely to serve violent roles accept feminism, statistically speaking. Why? Because the rich, coastal people have forgotten combat as a cultural frame of reference. The men of rich, coastal America don't have to do anything which a woman couldn't do, so they ask themselves why gender roles exist at all. Imagine the society opposite of that, and you have envisioned my father's country. "Quickly!

Smash the batteries open and pour their acid into this bottle." "Hide on the west side so the sun is in their eyes when they come."

The experience of violence informed every judgment we made about people and life and right and wrong. If a woman has a baby out of wedlock in my father's country, that child will not have any protection. Is the mom going to fight off a band of thieves? No. Without a father, that child will grow up in poverty and danger. That is unacceptable. Her orgasms are not worth more than the fate that child would suffer. If a man falls in love with another man in my father's country, that means at least two women have no protection from the world, and how dare they ignore their duty as men. That's not "intolerance." If my father rose up from the dead to see what I was doing with Kyle, he would compare the pain of a fatherless child to the benefit of an orgasm. My father would see any act of voluntary premarital or extramarital sex as me leaving my baby in the wilderness, hungry and alone. Bullwhipping might be too good for someone who does that. I'm not defending his thinking. I'm not saying that's the right way to think about things. I'm simply saying that's how he thought about it, and his mindset was rooted in experience, not in evil.

Every night of my childhood between the day my dad died and the day I decided to be an American, I dreamed of my dad's face on the desecrated half-corpse of Muzafar Avasov. I recognize that I should spend more time mourning my father's death than I should the photo of Muzafar Avazov, but I can't. When I think of my dad, I think of that photo. I'm not upset at my dad. I'm not telling you the story of how I was traumatized by a spanking. I'm telling you that the reality of my father's country drove my dad, who loved me, to force me to look at the corpse of a man who was boiled by his own country, and that experiences like that inform an ancient moral code which the globalist West has forgotten.

It wasn't cruelty. It wasn't abuse. It was an earnest act of love. That's why I want to forget it. And in the process of forgetting that country, I collapse and distort my father's memory. My brother has not distorted our father's memory. Perhaps that is why I can still tell stories like this one accurately, without making our father fiction-evil. I want to process that memory by creating distance, but my brother processes his memories by clinging to them, wrapping himself up in the beliefs of our father, and pronouncing words like our father did.

On top of him, I was making fun of Kyle for eating too much meat when someone banged on my dorm door. "ILNARA! OPEN THE DOOR!" the voice bellowed in Tatar.

I gasped. "Kyle, put on pants!"

Kyle floundered out of the bed like a walrus. I flopped behind him, entangled in all my pillows, falling down on the ground before scrambling up to put on my clothes.

Ilsur was banging and kicking the door. I knew that if I acted apologetically he would keep going, so I puffed up like a rooster and put my hand on the door handle.

"OPEN THE DOOR!" Ilsur roared.

"NOW, YOU LISTEN TO ME!" I fought back, opening the door. "This was not your decision!"

"AM I THE ONLY ONE WITH A MORAL BACKBONE IN OUR FAMILY? NOW YOU?" This time, he really did sound exactly like our dad.

I was stunned.

Ilsur pushed past me and stormed up to Kyle, who cowered behind me. "YOU DOG! YOU PUTRID DOG!" Ilsur shouted, arms raised.

I jumped on Ilsur's back and covered his face with my forearms, riding him on his shoulders as he tried to shake me off. "What is WRONG with YOU?!" Ilsur hissed.

"Calm down! Ilsur, I'm sorry. Promise me you'll calm down!" I wrestled, still on him, as I tried to kick him in the back of his knees. "Choose your sister or your principles, not both!"

"PRINCCCIPLESS? That's quite a WORD! If Dad was around, you'd be—"

"Dad's not around!"

"—halfway dead from shame!"

"I'm not ashamed, Ilsur!"

"*WHORE!*"

I shouted that he was an infidel with the Arabic word *kafir* and screamed in Tatar, "In order to be this vengeful you must believe that you are The Judge!"

He didn't acquiesce as much as he went soft. His muscles relaxed as he knelt down slowly. He barely opened his mouth. "I thought that I could trust you."

I started to cry. "No! Ilsur, no!"

Kyle rocked from side-to-side, frantic. But Ilsur was calm now, so I looked at Kyle and told him, "It's okay for you to leave."

Kyle didn't move at first. Then, he said, "Oop! All right," and scurried out with his tail between his legs. Kyle knew he wouldn't have won that fight.

My residents were curiously waiting outside, ready to call the police, and I said to them, "Everything is alright. Please just let me handle this. I feel safe." I did not feel safe, but I shut the door on the hallway of curious spectators anyway.

Ilsur whispered again. "But why?"

"Did Haleema tell you?"

"Why?" Ilsur barely breathed.

I was about to say, "because it's my body and my life and I'm an American," but my brother made direct eye contact with me, and I saw true rage. I hadn't felt ashamed before, but his gaze broke my American psyche. His irises were so dark, it was hard to distinguish them from the pupils in this light. I imagined leaving my baby in the wilderness.

"Hmm," Ilsur managed to utter. Finally, he walked out with his shoulders hunched, his chin hanging over his chest lower than ever, forcing his feet to walk away like they were going to turn around without his say. He didn't say anything. He just walked out.

I shouted at Ilsur's back, "I told you! It's your sister or your principles!" as I started to sob to myself.

I looked at my phone. A text from Haleema. "Your brother is nuts. You might want to talk to Ilsur soon."

Alone again, I texted Kyle that I wanted to sleep over that night. He said, "I mean, we can't sleep together every night, on principle." But what he meant by this was *I am scared of your brother, and I care about myself more than you.*

"Kyle, I need you now," I replied, feeling more alone than ever.

He texted, "No, I just really think we need distance right now—to think through things on our own."

I decided to go out and find Kyle. I looked at his Snap location and saw he had gone to a bar. He must have been drinking the distress away, so I called an Uber in my pajamas. The bar smelled like old wood. It wasn't a college bar. I looked around. Indeed, Kyle had gotten very drunk, and a random blonde woman on his lap laughed as she twirled his curly hair.

I didn't confront him. I just thought one word, which rang in my head in a baritone voice and Russian accent.

Despicable.

The dormitories bent sideways through the refraction of the water in my eyes on the way home. I shut the door of my bedroom and laid on the floor, like people used to before they had beds. I reimagined what happened with Ilsur. I had half-expected my brother, as he walked out of my room, to spit on the ground. In that moment, I hated myself. I was sorry. I was so deeply sorry. I was also free to do as I please. But I had been chasing this dream because Americans are so much happier, right? Right?

Because no one prepared me for what Kyle did.

No sitcom.

No conversation with my peers or professors.

Only Ilsur.

I desperately wanted Ilsur to have spit on the ground so I could have screamed at him for hours. I wanted to bash his head in, and I wanted him to tell me he still loved me.

He probably wanted the same thing.

I didn't see it.

Dad fought to cast off Soviet bans on religion. We celebrated the attacks on the capital in Tashkent in 2003 because the little bitches of Russia that we called the Uzbek government got what was coming to them. The government disagreed. The uniforms got orders to fight extremism and they roamed the country tactically, taking out the leaders of Islamic revivalism.

The way my mother's friend told it to me, they captured my dad in a meeting. Took the men from the meeting outside. Lined them up, put a Tokarev-38 to the right temple, and shot.

I laid in my bed wondering if I should go to Ilsur's apartment to talk it out, but there would be no talking. I imagined myself naked from the waist up, arms tied around a post, the skin on my shoulders shredded up and dripping blood. I know that mental image shouldn't be comforting, but there is peace in knowing that you got what you deserved. I wanted to call my mother, but she would just resent Ilsur more. My mom didn't take judgments like that anymore. She was too old and had already made her choices.

No, my primary problem wasn't that Ilsur was controlling or overprotective, although he was. It was that, however controlling or overprotective, there was something very right about what he said, and very horrible about my actions. Kyle didn't have the feeling to fight for me. Kyle didn't have the sense of duty to stick around to make sure I was safe afterward. Kyle didn't even consider fidelity important. He hightailed it for another woman when things got bad. It took him all of two hours. *That's what happens when you hook up like an American,* I thought to myself.

I felt cheated. I felt gross. None of my orgasms were worth the pain of realizing his utter impotence, that I had shared my soul with a half-man. I was losing Ilsur for ... that?

I had to apologize to Ilsur.

I had to.

But it was my decision! And my body! And he doesn't get to make these decisions.

I turned on my belly and remembered that Hanif Kurieshi wrote *My Son the Fanatic* about a boy like my brother. In the story, the family moved to England. The father found a white mistress, so the son found God. There is a young generation in the colonized world like that. Like Ilsur. Students fighting for the former glory of their country. Students finding God and demanding authentic practice of the customs they used to have.

To go forward, they go back, just like Nietzsche said.

I laid the small pink blanket on the floor in the direction of Mecca. I washed my hands, sucked water and spit it out, washed my nose, my face, my arms, behind my ears, and my bare feet. I draped a sheet around my head to cover my body. I stood on the far side of the pink blanket and recited *ishhadu anna la allihe ila Allah* in a Russian accent, standing, bending, declaring *God hears those who call out to him,* and standing. Does God hear those who call out to him? Because I'm calling. I'm sorry.

I kowtowed, raising up and then touching my forehead to the ground, *God is greatest.* Again, I kowtowed, raising up and touching my forehead to the ground, *God is greatest.* When I was curled up in a little ball, while the prayer was coming out of my mouth, I saw an orb of light in the back of my eyelids. I asked it to tell my dad that I missed him and that I was confused and that I didn't know what to do. I asked it to tell him that I'm sorry. By now, we had learned that Ilsur wouldn't do anything violent to Kyle, but I had no idea if Ilsur would still want me in his life. I had cast off the knowledge of my former country, but maybe that knowledge was telling me something important. Kyle broke my heart after our first fight—a fight I didn't even start, a fight I took his side on. It took him two hours. Two. Hours. *I want my family more, God.* Ilsur might cut me off, cut out those pieces of his heart which loved me. And if he cut me off, he would be almost completely alone. *Please*, God, *keep my brother in my life. Please.* I greeted the angel on my right shoulder, *peace, God's mercy and blessings upon you.* I greeted the angel on my left. *Peace, God's mercy and blessings upon you.* The light turned red and yellow and white. *There is no god but God. Peace upon you.* I rose up and blinked, realizing that it was just my lamp through my eyelids.

Early the next morning, I heard a *ppsss* under my door. I hadn't slept, but I couldn't bring myself to get up and look at the paper. I turned over and covered my face with a pillow.

But the curiosity tickled me awake eventually, so I swung out of my dorm bed and looked at the paper. It was loose notebook paper with frilled ridges on the left side from the notebook wire. In chicken scratch, the paper had the last three Microeconomic homework problems I had asked Ilsur to help me with written out in full. Yes, God hears those who call out to him. I took the papers close to my chest.

I love my brother, I do.

"Black lipstick?"

"That's the mood," Diep agreed on the day we met. We both ended up at the dormitory hangout because we didn't have anything else to do that Thursday, and I naturally approached the thin girl with black lipstick.

"I think black lipstick is a rejection of the way people look in the wild, and therefore an attempt at ascendence towards non-human forms," I said in the characteristically awkward way I quickly move the conversation to something interesting.

Diep looked at me and understood. "Like a rejection of our nature?"

"Yes, exactly." I was so relieved I had found someone to talk to.

"I mean, any makeup is a rejection of natural beauty," Diep observed.

"Well, I wouldn't say that." I leaned in. "I think that most beauty products are designed to enhance natural beauty: flushed cheeks, long eyelashes, clear skin. But purple hair and black lipstick reject the human color pallet, right?"

"No," she disagreed, "because you're not enhancing your natural color when you wear red lipstick. People know you're wearing red lipstick. That's why you wear it. You're signaling to the world that you're put together. Mostly to other women."

"Mostly to other women?" I raised my eyebrow and made eye contact for the first time. Diep's irises verged on black, situated under epicanthic folds. From her face alone I didn't know if she was South Asian or South American. I couldn't ask her because at Morrison they teach us it's white-normative to acknowledge other people's races.

"You know," Diep laughed awkwardly, "other women know when you're wearing makeup. It's a performance for them."

"That's one reason I don't wear makeup." I admitted to her.

"Yup, and that's the reason I wear black lipstick."

Diep invited me over to her place that Saturday.

I walked through the door and saw no roommates. "You live alone?"

"No, I technically have a roommate but she decided to live with her boyfriend." I didn't think anything of it. "You can crash in her bed if we get too lit."

"You're throwing a party?"

"No, but I figured we could drink." Diep poured the Barefoot bubbly into plastic cups and I took one.

"I like what you've done with the place. It looks really good for standard housing." Diep had meticulously ornamented her apartment with yellow, red, and pink mosaic fabric and scented candles.

"Yeah, it's hard to make their dentist office couches look good, but I figured it out," Diep complemented herself.

"It's always interesting how people are able to transform their space," I observed as I sipped.

Diep bashfully changed the subject, "Yeah, I like that stuff. It's why I like watching *What Not to Wear*."

"Oh, that's the best reality TV by far," I agreed.

Drinking too much, we watched an episode together, then watched halfway through the second episode, stopping when London and Clinton criticize the woman's clothing as she stands in the mirror. Diep interrupted, "Do you want to watch *Criminal Minds* instead?"

"I mean, not really. I don't like scary movies," I lied.

"It's actually not scary because it's formulaic." Diep explained. "The scariness of horror stories happens because you don't expect what's coming, but once you watch one episode of *Criminal Minds* you can predict every scene in every episode thereafter. It's just fun." Diep understood how stories worked.

*"Thereafter* is a nice word. It's so old and formal." I observed, distracted. I didn't want to watch Criminal Minds because of something I couldn't tell Diep about.

My first schoolgirl crush was not any boy I knew—not even a real human. It was Spencer Reid from Criminal Minds.

When I was eleven, my parents split and my mom couldn't come home early, so she got me a television, a computer, and a phone the same year my father no longer occupied a presence in my day-to-day life, as divorce always does, even though both of my parents loved me.

In sixth grade I started watching Criminal Minds. I fell in love with Spencer. When I played *The Sims* with my cousin, a videogame that let us roleplay our lives, I kept designing variations of Spencer, coming up with stories about how I would act if I met him. I would play it out on the computer for hours. I put Spencer into Harry Potter, rewriting Rowling's plot in my head to somehow make the character relevant; scrapping the character Hermione so that Spencer and I could fall in love in Harry Potter's universe. The video game let me daydream like a child with the sexual intensity of an adolescent.

Surfing the internet, I realized that some Criminal Minds fanfiction was written to be arousing. Terrible writing, yes, but nonetheless provoking. I felt disgusted by the sight of actual bodies. Porn videos are so quick and catered to men. Written erotica, however, fleshed out the emotional journey of falling in love and ended with sex. By twelve, I was consuming porn stories about two hours every day, sometimes exclusively about Spencer Reid from Criminal Minds. I loved Spencer. I really loved him. I read so many stories and backstories and imagined scenes about Spencer that I could almost convince myself that he was a real person. That mental state of temporarily forgetting who is real, convincing myself that what I was reading was real because it is so detailed and complete; that is the danger of the internet. Reality gets lost.

So Diep and I watched Criminal Minds together, and I noticed that Diep laughed at the show at the same times I did. Diep crossed her legs and bent her hips forward when Spencer said something funny, clearly putting pressure between her legs. I experienced an emotion that could be mistaken for sapphic: Diep was aroused by the same television character as me, so I felt a sexual connection with her. Two voyeurs staring into Spencer's life.

My roommate and I didn't click. Diep had an extra bed. I spent so much time at Diep's apartment that I basically became her roommate. One night around 4:00 AM her boyfriend knocked on her door.

"YOU'RE A CHEATER!" Mabior Jok banged and screamed from the hallway. Diep flicked on the light and went towards the door.

"Are you crazy? Don't open that!"

"No, I need to make this right." Diep humbly walked towards the door like she didn't understand the imminent danger—as if she wanted to get the shit beat out of her, which she did. She had never experienced pain ever. Never once. Any time she was hurt, she took a pain killer. No one ever intentionally hurt her. Because she had no memory of pain, the idea was alluring to her. She endangered me and herself by opening the door, "Mabior, listen, I was really drunk, and he kissed me first. And I'm sorry."

"Give me your phone," Mabior demanded. I stayed curled up on my bed with a pillow over my eyes, resolving that I would go out there only if I heard him throw a punch.

Diep handed him her phone. He deleted some numbers from her contacts and blocked the guys on her Snapchat.

"You're disgusting, you know that?" Mabior kept repeating. "Absolutely disgusting."

A couple weeks later over lunch in the cafeteria that the doctors at Morrison Hospital use, Diep told me, "When we were at the mall I wanted to get my eyebrows threaded. They had a sign that said $4 off. Mabior also wanted his eyebrows threaded, but they said the deal was only for women. He screamed at them that that's not what their sign said, and argued and argued until they just gave in and gave him the discount."

"How sweet of him." I took a bite of scone. "You know that's not normal, right?"

Diep changed the subject. "I don't want to talk about Mabior. In less than an hour I have to go convince my actuarial mathematics professor to not ruin my life."

"What happened?"

"She just won't budge." Diep explained, "I want to be an actuary. She can't take that dream away from me."

"What isn't she budging from?" I asked tentatively.

"She didn't give me accommodation for my depression, so I didn't pass any of the tests."

I paused. "What type of accommodation did you need?"

"She made me take the test on the exact same day as everyone else even though I wasn't ready because I was receiving treatment." Diep snapped, "She couldn't work with me on my serious medical issue, and that's discrimination. She's ruining my future because I have a disorder." Diep stuffed her nose into her phone, logging on to her other Instagram accounts so she could get her 'like' count up. She had posted a picture of us together that wasn't flattering of me, and it made me upset.

"That's terrible," I agreed with her because I didn't want to lose the relationship we had built. But I knew it wasn't discrimination. Diep didn't study and therefore she did not deserve to pass. If her disorder makes her incompetent for a job, then she shouldn't do the job. If she tried to do a job she wasn't qualified for, it would hurt the business and the clients. But I just couldn't say that out loud. I didn't want to be accused of discrimination. Also, I knew that she didn't understand. She genuinely believed that the professor was being mean.

That night I texted Diep, "How'd the conversation go?" but I didn't hear back. I didn't hear back the next day either. I didn't hear from her the day after that.

On the fourth day I got a text message saying, "Sorry, I was in the hospital."

"Goodness, are you alright?"

"That professor ruined my life and it just wasn't worth living anymore." Diep texted. And I swallowed a rock realizing that she had attempted suicide. But three days later she got out and resumed her normal life.

"I want your twentieth birthday to be fun!" Diep told me. "We can go to the noodle bar in Morrison village." She invited my seven closest friends from my sophomore year of college. But she took too many pills and fell asleep and didn't come to the birthday party she had organized until it was over.

I tried to confront her. "Where were you?"

"I'm really sorry. Like, I should be a better friend than that and I'm sorry."

It seemed like she understood what she had done, so I accepted her apology.

After I finished with my finals, I told Diep that I would help her move into her summer home. As I cleaned her bathroom, I noticed full pill containers upon pill containers. I stuffed them into a box to help her move. The label for the little pills said to take three tablets before bed, and more as needed, and I thought that seemed like an awful lot.

As we unpacked in her new apartment, I asked her, "How many credits of summer school are you taking?"

"As many as I can. I have a lot to make up. Hang on, my mom is calling." Diep answered the phone. I expected her to speak in Vietnamese, but she said in English, "Ugh, mom, I'm with my friend right now." She listened. Then she snapped, "Honestly, no, I don't want to hear about your yoga retreat. I'm with my friend right now," and hung up the phone.

I knew it was a bad idea for her to take too many classes in the summer but I didn't say anything. If I had said something, it might have seemed rude.

Diep noticed the silence and asked, "What are you doing this summer?"

"I just have to go home and rest," I said. I had a lot to work through the summer after my sophomore year.

It's hard to explain what I needed to process that summer.

I knew that I was very accoladed at the things I was good at: writing, foreign language, pleasing teachers. I had published two pieces, received a coveted fellowship, got into a very good school, blah blah blah, but I could not tell myself apart from the years of anger which hung in my throat. At 20, someone might say something which would remind me of a memory. When the memory came, it transported me away from this world to somewhere under the ocean; where I was embarrassed, alone, guilty, and pathetic.

That time I couldn't remember.

That time I told a joke that fell flat.

Hard times.

Times people hurt me.

Times I hurt people.

The flashbacks weren't the worst of it. Sure, fork prongs reminded me of syringes, cancer, baldness, ugliness, pity. Sure, a kid picking his ears reminded me of Shrek and how I felt in my body when I was fourteen. The flashbacks sucked. But the worst was that I could not properly define myself. I was not a series of adjectives. I was not from a definitive background or a member of a cohesive, united group. Other people were more successful. If another person was not successful, it seemed like at least they had external factors to blame their life on. I did not have that. I was responsible for my problems, and my problems sucked, so I sucked. The things I hated about myself were me. My history. My decisions. My mind. That's what I hated.

When I say cancer, I mean both the time I underwent treatment and the way my body remembered it. The weird shivering of my limbs. The weight gain. The compulsion to watch my blood drip instead of getting a Band-Aid because my arm was a thousand miles away from me, like looking at someone else's arm. Overeating. Rubbing alcohol smelling good. My cancer was of the immune system, so I was kept in isolation for years of my childhood. I talked to myself. I still talk to myself for hours. I have engaging, irritating, phenomenally grand conversations with myself when I am alone. When I had cancer, my parents were together as a family. In a way, cancer was my family. It was what kept my family a family. Not long after I went into remission, my parents split.

Internally, I had no idea who I was. Even before Morrison, when I went to public high school with mostly minority students, if you count Arabs and white Latinos minorities, I was allotted the labels "white" and "rich". But when I came to Morrison, the gap between my whiteness and richness was so, so far below my peers that I felt estranged. Diep liked to say that she was a victim of the master's gaze because when she wrote stories as a kid they were always white characters. The effects of colonialism are real, but Diep had the master's gaze more than anyone I knew. Objectively, I didn't. When you go places like Morrison, universities for the children of the rich, you meet people who vacation in Europe and teach their children Mandarin and talk about Cyrano. Rich people think macchiato means a tiny cappuccino, but heartland Americans think macchiato means caramel in your coffee.

But I was not a low-income student or first generation, either. When I sat in some English classrooms, I found myself feeling ashamed that I wasn't low-income enough! It felt like, after reading Audrey Lorde, the person who oppressed was the bad guy, so the person who faced oppression was the good guy, and the person who faced the most oppression was the best guy, like an inverted pyramid. I felt terrible guilt for having a mother who paid for my college. As I spoke to my mom about it, she didn't understand. She just couldn't understand the estrangement I was feeling. "Let me get this straight, you're crying at me on the phone because you *won't* graduate with debt?" I realized that feeling ashamed about being middle class only happens in the weird milieu of elite universities. I cannot think of a single person from my high school, or my cousins' state college, who sits around feeling proud about being poor. When I first heard someone invoke their financial aid as a way to bolster their legitimacy in a club meeting, I was confused and began questioning myself instead of feeling grateful. I was on the borderlands between the identities people taped on my forehead without my say. It confused me.

And then there was Taras. He's half Uzbek. What scares me about Taras is not his affinity for weapons, violence, the alt-right movement on 4chan, or his stunted sense of humor. No, Taras scared me because I loved him. It felt like floating. When I was in love with this man, having sex, the world with all my rage, insecurities, and loneliness melted away. He filled my heart. He took care of me. He didn't feel guilt for acting like a man. He didn't care if people disliked him. When I was around him, I could be me, and I sponged up his strength enough to not feel bad for being weird and fat and standoffish. He absolved my insecurities by being okay with his. Poof, gone. I loved him enough to cover every one of his flaws with compassion, and I didn't care about how much he hurt me.

Taras had a funny tendency to bring up my insecurities to win an argument. "Of course you would say that, you're not a very fast reader. Not really quick on the uptake of new ideas now, are you?" He told me. Once, after I screamed at him for not listening to me, he pinned me down and forced himself into me. Just grabbed my thighs and shoved them open, laid his elbow on my throat, forearm across my face. Another time he slapped me in the face so hard my head hit the window of his dad's truck. Another time, when he was losing a political argument to me, he told me he'd wish I'd cut myself.

And you know what? I was as mean to him as he was to me. I surprised myself with my own cruelty towards Taras. The worst thing about being with him was how I compromised and lost my sense of kindness towards my fellow man. I had never screamed at, slapped, or intentionally hurt anyone in my conscious memory before then—I had hurt people before, but not intentionally until Taras. A week after he shoved himself into me without my consent, he was arguing and I looked him straight in the eye and said, "This is why your mother never loved you." That comment tore his soul out of his body, and I watched him writhe in pain. Taras had little teeth and sunken eyes, and when I think about him,

I imagine his teeth growing longer and his eyes sinking back in his head.

In our last interaction, Taras and I watched a solar eclipse. The eclipse was incredible. The land grew dark and in the sky I saw the eye of a cat. I dumped him but I missed him. He hacked my computer. I threatened to call the police and dumped him again. I blocked him on my phone. He rerouted his cell phone number through India to get past caller ID. I dumped him again. He showed up at my grandma's doorstep on Christmas. I threatened to press charges and haven't heard from him since.

So the summer after my sophomore year I needed to rest. I needed a break. But I didn't have any other female friends in Lansing near my grandma's house, so I needed Diep.

I flew back down to Houston to visit Diep in the summer. Diep lived in a royal apartment building with a waterfall fountain running into a heated pool. Diep invited me and her friend Erica over for margaritas by the pool the day I arrived.

"Where's your roommate?" I asked as I put my suitcase on the granite countertop kitchen island.

"She decided to move out. I don't really know what's up with her." Diep explained. Diep's phone rang. "Mom? Honestly, no. What?" Diep hung up.

Even though Diep spoke to her mom like I would to someone who cut me in line at Walmart, I assumed there was a dynamic there I had missed. Maybe her mom called her all the time? Whatever, I was Diep's guest, so I didn't say anything out of respect. I slipped on my bathing suit and asked, "So how do you know Erica?"

"She's another pre-biz student in summer school with me, on the actuarial track. She's been studying with Mabior and me. Helped me with a test on probability."

"I thought Mabior graduated?"

"Ya, he did, but he's sticking around for the summer for an internship." Diep explained. Then she turned around and took out a bottle of sheer little black pills, took three.

"Three?" I asked.

"Um, ya," Diep responded as if to say, 'mind your own business.'

"Alright." I shrugged, flinging my towel over my shoulder. "What's her last name?" I switched the conversation back.

"Erica Shen."

I whipped out my phone and noticed her posts got almost twice as many 'likes' as mine on Instagram. They were almost all her standing in front of plants.

We met Erica poolside. Erica looked pudgy in her one-piece swimsuit. Diep looked like a short barbie doll in a bathing suit which cost $200. The swimsuit annoyed me. What a waste of money, I said to myself to cover up the seething jealousy of how good she looked in it.

I was thin by then, which elevated my social class and sexual capital, but I hadn't shaved in a while, and I've never been present enough to have good style. My focus had been on my mind instead of my body. An embarrassing patch of pubic hair peeked out of my bikini. I looked around to see if there were any guys, but none were close by, so I let my towel down and sat on the edge of the pool, swinging my feet in the water.

Erica noticed my pubic hair. I rolled my eyes. I hated having to play this game where she knew I knew she knew but she didn't want to make me feel weird. I blasted through the awkwardness, "Oh, yeah, shaving gives me razor burn, and I don't want razor burn down there, you know?"

Erica nodded, a little embarrassed she was caught staring at me there. She asked, "Can't you just use shaving cream?"

"I do. I swear every time I shave, I get a sea of ingrown hairs. It's just so gross I can't do it."

"Can you cut it?" Erica suggested.

Diep giggled, "Get a guy's beard trimmer!" and we all started laughing.

Erica and I enjoyed each other's company, so we decided that night the three of us would hit downtown Houston and visit some techno bars.

"Just one dance?"

"Oh my, my, my."

"For you? I'll buy Chinaco."

Diep took out her iPhone and posted to her Snap story. Then she leaned in and asked the man in front of her to buy her a drink. She slurred her words and snapped more selfies for her story. I don't like loud music so I wore two sets of earplugs, but the men didn't notice. Guy after guy threw themselves at Diep and me as we danced, but they ignored Erica, which created a dynamic of resentment and embarrassment. This was completely unfair, so I stopped dancing. Diep continued.

Diep is small, beautifully proportioned, and ethnically ambiguous—is she Latina? Native American? She attracted many different men. Eventually Erica leaned in and said to me, "Let's count how many languages Diep gets hit on in a single night," and that's when clubbing became really fun.

As Erica and I were roaring with laughter, out of the corner of my eye, I saw an angry face with properly threaded eyebrows. Mabior. He had seen her snap story, followed her to the club, and watched man after man throw themselves at his girlfriend. And he watched her use those men to get them to buy us drinks.

Mabior waited until Diep also saw him. She froze in her tracks on the floor. Mabior didn't do anything. Just stared at the three of us, then turned around and walked away.

The next day, Diep received an email from the Morrison Honor Council. It said, "You have been accused of cheating on a test in Actuarial Mathematics 252 for the chapter on probability. You will have the opportunity to speak your case on August 22nd, 2018. Morrison University considers cheating a serious matter."

Mabior did it. Mabior had graduated. Even though he had knowingly helped Diep cheat, he wasn't a student anymore so there's nothing the school could do. Mabior didn't accuse Erica. But the honor council could get Diep, and they did. She failed the course, got a 2-year mark on her record, and left the actuarial mathematics program. Mabior won. Diep lost. I knew that Diep had been stupid and selfish, and that she had been cheating on her man, which is wrong, but she needed someone on her side. She was just lost, you know? Mabior should have just dumped her. He didn't need to be vindictive like that. He's not innocent. When I looked at Mabior, I saw tiny teeth growing longer and his eyes sinking back.

The night Diep came home from honor council, covering her face with her hands from the shame, she got her smartphone and opened Tinder. She wanted to show Mabior whether he owned her once and for all. She ended up in the hospital on a course of anti-AIDS medication, drowning herself in big gulps of Plan B.

In October of my junior year of college, I ran into Erica, who apparently was part of the same RA legion I had been hired to work. I asked Erica if she had been keeping up with Diep.

Erica slanted her eyebrows, "You don't know what happened?"

"You mean with Tinder?" I clarified.

"No, after that. Once Diep got out of the hospital, I wanted to pick her up, so we went to Miami for the weekend. She was still on the anti-AIDS medication, you know. I told her, 'let's just have a good time,' but she really wanted to go out, so we went to a bar. She texted me that she was going to kill herself, and I looked over and saw she had ordered fifteen shots, downing them all by herself. The bartender didn't realize they were all for her, so he had to yell at her and we rushed her to the hospital."

Erica hadn't talked to Diep since. Diep needed a good friend now more than ever.

In the beginning of my junior year of college, I went to Diep's apartment and saw that she had three roommates, one named Puteri. "Nice to meet you," Puteri smiled and offered me a brownie.

"I'm sorry, I'm trying to cut back on sugar." I responded.

Diep went into her bathroom and came back with a scarf. "This is yours. I smelled it and thought, 'yup, that's Jane-Marie's'."

"Gross." I took my scarf.

"You don't need to be judgmental like that at yourself," Diep said with out-of-character language. "You just smell a specific way, Jane-Marie. You don't need to attach judgment to it. It's neither good nor bad. That's what my counselor and I have been talking about."

"You don't think it's gross to smell like body odor?" I asked.

"It's not body odor. It's the scent of your body, which you can't change so you need to accept. Like, I have depression that I can't change, so I need to accept it."

Diep promised me that she would read my story for a class. "Of course I can look over it," she assured me.

She never read my story.

"I can't go all the way back to Vietnam for Thanksgiving break. Can I come to your parent's place?" Diep asked me.

"I think so," I said. "It will be at my Grandma's house."

"But can we get vegetarian food?"

"Of course, my mom will make sure there's something for you to eat."

I called my Grandma and she said she was willing to make accommodations for Diep. That Thanksgiving, my grandma bought an entire vegan turkey substitute and learned how to make it, and she paid professional cleaners to come to her house because she was too old to clean like that herself. Everyone in my family was excited to meet my rich and successful friend from my rich and successful school.

Diep got distracted and inebriated and missed her plane to Michigan. It was okay, though, because she was rich and she would have embarrassed me in front of my family. I felt relieved when she missed her plane.

"Hey, do you want to meet at the library?" Diep asked.

"Yup." I walked over and sat next to her at the high tables.

"I'm going to write this paper, I'm going to do it," Diep said. She looked at her computer. She scrolled on Facebook. She posted on Snapchat. She got hot chocolate. She got coffee. She said, "Hey, do you want to study in a different area? I just can't study under these fluorescent lights. I feel like I'm in the torture chamber of *1984*."

At midnight, Diep emailed her professor asking for an extension. The professor apparently gave it to her. My professor in a similar course did not give me one.

"Hey, Jane-Marie, do you want to go out with me tonight?" Diep texted.

"I don't really want to go out," I said because I didn't want to deal with her drinking. "Maybe we could just have a fun night inside?"

"Sure, come over later," Diep texted.

I got to her apartment and she wasn't there. "I'm sorry, I just really wanted to go out. My counselor tells me that I need to keep putting myself first. But you can come meet me if you want."

Towards the end of the semester, Diep realized that she would flunk. She hadn't studied at all, but she expected her professors to pass her anyway because she could have done the work if she didn't have depression, she thought. The fact that she didn't do the work was beside the point. Her professors were supposed to take her word for it. Many, many times at Morrison, the professors accepted the path of least resistance and passed her through. But some sticklers wouldn't.

We were sitting in my favorite spot in the Spanish lobby on the fifth floor when she called her dad. It must have been midnight where he was. He advised Diep, "You need to seek medical leave from the university so that they understand your position and work with you. You know that your mother is very worried."

"I know she is." Diep spat, taciturn.

Diep requested medical leave withdrawal from the semester on the basis of a note from her counselor. She enrolled in a treatment program that would take months.

I ran into Diep's roommate Puteri on the bus to campus. "Hey, how are you?"

"You know." Puteri looked uncomfortable.

"Diep's not around, I know." I cut to the chase, assuming that would be the reason.

"I mean, I expected it. Everyone from our 'Women in STEM' program knows what happens with Diep."

"What do you mean?"

"I mean, she took too many antidepressants in the middle of the living room again. She did it before with Bailey. She never does it in her bedroom, you know? Always in public spaces where we'll find her. And if we don't do anything then we're at risk for manslaughter charges, so obviously, every time, we call and they take her away."

"Must have been hard to watch."

Puteri continued, "Yeah, she poured a couple pills from her antidepressant container into her hand and took them. I'm just sad that I wasn't popular enough to have other roommates. It's just like, I'm cool, you know? It sucks that people think of me as friends with Diep."

"No one thinks anything like that about you," I assured, but my mind was focused on something else. "How many pills did it seem like she took?"

"I don't know, five? Seven? The bottle wasn't full to begin with."

"They were the sheer black ones?"

"Yes."

Diep took large quantities of pills all the time anyway. In all of her suicides, Diep never tried to cut off her oxygen supply. She never tried to buy a gun. She took dosages of antidepressants that were large but, relative to the number of antidepressants she took anyway, a reasonable observer could infer that the dosage was not intended to be lethal. I swallowed a rock as I realized that Diep wasn't suicidal. Diep was in a lot of emotional pain, yes, and she was self-harming with medication, yes, but it is not honest to call what she was suffering from 'suicide.' She faked suicide. It was a performance for her.

And I realized that's why Diep always had that extra bed.

Diep enrolled in a three-month long program. Her friends from college Taylor, Sylvie, and I went to visit her once in the residential treatment program.

"Welcome to the crazy house!" Diep greeted us like it was a joke. I couldn't stand being there. I itched to leave.

"These are my friends Darcy, Emma, Ryan," Diep pointed at three other patients who were approximately our age. I couldn't help but notice that they were fat and their gums rode halfway down their teeth like they didn't floss. I wasn't better than them, and I hated that. This place showed me my ghosts.

Diep took us into the cafeteria, which smelled like Bosco sticks. There was no coffee, just water, because the patients would abuse caffeine. As a hairnet served my Salisbury steak, a hamburger patty in some gravy, Diep roamed the tables to find a place to sit. She looked like a freak of a butterfly among moths.

I uncomfortably sat next to Ryan, who noticed my discomfort and blasted through the awkwardness by explaining, "I'm a musician. I got caught up in drugs."

"A lot of musicians do." I tried to empathize. I noticed a round scar on his eyebrow that looked like the indentation of a ring.

"It's hard not to when everyone around you is doing it," he explained.

"Absolutely," I agreed. I knew that wasn't the whole story because this was a psyche and trauma rehabilitation program. His scar seemed neither fresh nor self-inflicted, but I didn't want to pry.

"So, how are you Jane-Marie, Taylor, Sylvia?" Diep asked us. We gave our updates.

I said, "I've been having trouble in the journalism club with Jessie."

Diep advised, "Well, you have to remember that you are in college to get a degree, not to be in a club. Focus on academics. Graduate with honors, you know? You have to look at the big picture and focus on the things you can control."

My muscles relaxed and I smiled, relieved. "Yeah, you're right," I agreed. It was as if she had made the entire situation disappear from my life. Diep made me feel better. She supported me. She was capable of empathizing enough to make me feel better, which meant the treatment was working. She was getting better.

A month or so after that, the situation with Jessie in the journalism club did not get better, and I wanted to talk to Diep about it. When Diep was a level-three patient she could leave the program for a couple hours per day. When we met, I asked, "Do you feel weird at all about the fact that other people in that program are, you know, like, poor?"

"Not at all, why would I make those kinds of judgments?" She asked me confused.

"I don't know. It's just, like, can you tell they think you're different?" Diep had always been too entitled and didn't understand that caused her social problems. I believed that while receiving the treatment she needed, the doctors would have gotten her to a point that she recognized her entitlement.

"They don't think I'm different." Diep cut the air.

"I mean, when Sylvie visited your family's house in Vietnam, there was a servant who stood in the corner of your dining room to fill up the cups of tea."

"That doesn't mean Ryan thinks I'm different. We actually started dating!"

"Oh, that's nice." I decided to let it go.

Then I tried to talk to her about what happened in the editing club, and she said, "I'm not your therapist, I can't help you on that."

Diep was angry, so I left early.

Diep got out of the treatment program and stayed with Ryan on-and-off. Except Diep went on Tinder and she found a strange, old, gross man who I call Mountain Dew Ronny. Sylvie texted me, "I'm going to cut her off. Taylor is, too. She just has no self-respect."

The next time I hung out with Diep, she asked me, "I can't tell if Taylor and Sylvie are actually angry at me or if that's just my depression telling me they're treating me differently."

"Okay, well, I'm going to tell you the truth because I think you deserve to hear it. They are absolutely livid at you. You have to choose between Taylor and Sylvie or Ronny, because you cannot have both." Because sometimes you owe it to the people who love you to keep it together.

Diep looked down, then concluded, "I'm going to fight for these friendships."

Diep cut Ronny off. She was getting better.

I invited Diep over to my place, a tiny room in a sorority house with paper thin walls. I was the RA of the sorority house, which is kind of like being a tax collector in Judea. Diep spent the whole time on the phone with an insufferable mutual acquaintance named Alex who claimed to be suicidal like Diep. I knew Alex. Alex judged people very harshly, but she could not figure out why she judged herself so harshly. If you constantly judge and berate the people around you, eventually that malice will turn inward and you will judge and berate yourself. Alex did not know that, so Alex couldn't resolve her pain. Diep called Alex's mom and warned, "You need to go over and help Alex." We started talking about Alex.

"I wouldn't have done that. I respect her right to make her own decisions," I said. If Alex was really suicidal, would she have spent three hours telling Diep how sad she was over the phone? I was talking about Alex, but really I was talking about Diep.

Diep got the hint. "No suicidal person is in the right state of mind to make the decision to die, Jane-Marie."

That line shut me up, because I knew that if I said anything else, I would be misconstrued as the bad guy. You know, like Canada and its Medical Assistance in Dying. I was already the bad guy, actually, because of what I implied. So I stopped talking. I didn't have the framework or vocabulary to defend my point because I couldn't bring myself to say that rich assholes like Diep and Alex who claim to be suicidal often clearly do not actually want to die. They don't need better self-esteem. They don't need less self-judgment. They drown in their own egos because they don't have any obligations to anyone else: zero obligations, zero responsibility, zero purpose, zero need to exist. I know it's an old adage about other people having it worse, but seriously take time to consider that many millions of people pull rikshaws for a living—they're literally doing the jobs of mules—and judging by attempted suicide rates, those people have more purpose, meaning, and self-respect than rich kids in Western colleges. I recognize that Diep and probably also Alex experienced real emotional pain, but who the hell hasn't?

My sense of shame blocked that sentence like a weight on my throat. It just wouldn't come out. It was so obvious that Diep was full of shit, but the moment she wrapped me in a rhetoric ring, I felt an immense social pressure to support her, help her, and believe her even though I knew she was lying.

And you know what I realized? The doctors responsible for treating Diep must have felt the same way, as evidenced by their lack of success in treating her. If they were competent experts in mental health, then they probably knew Diep's case profile differed from that of a person experiencing an earnest crisis. Yet, the hospital had been treating Diep as if she was experiencing an earnest crisis, goading her to develop better self-esteem. That probably had something to do with social pressure on doctors in the hospital.

Taylor and Sylvie had moved home during COVID. Diep and I continued to hang out together in quarantine. Taylor and Sylvie's relationship with Diep stayed intact. I hung out with Ryan and Diep and my boyfriend.

Diep and I went out for lunch at a restaurant in Galveston during COVID. At the end of lunch, Diep asked me to get my eyebrows
threaded with her. "I know a really good place. They wear masks; it's
safe."

I took out my phone and looked at the pricing. "Can we go somewhere else?"

"I'm not comfortable going anywhere else."

"Seriously? You literally just ate at a restaurant during quarantine," I
snapped.

"Outside. I ate at a restaurant *outside*."

"You literally went to Galveston beach without a mask last week." I
reminded her defiantly. "And you're going to have to get your eyebrows
waxed inside any place we go."

"Which is why I only would feel comfortable going to this one
place!" Diep shouted as she stared me down. As if I dare defy her.

"Fine, alright we can go there," I acquiesced. But it was so stupid.
Get this—Diep might have felt *uncomfortable*. You know that feeling when
you sit in a chair with poor lumbar support? When you're not entirely
comfy? That's what Diep might have had to feel. Good God, the horror.

My phone rang in the middle of the night. It was Diep. "What's
wrong?" I asked sleepily.

"My mom is fat shaming me," Diep sobbed.

"What did she say?"

"She said that I had to start seeing a personal trainer because I don't
look healthy right now. It's just because she's been talking with her stupid
fucking friends about who I'm going to marry. All of those women are
just assholes. It's like, it's my body! Who are you to judge my body, and
decide for me, and. . . fuck you if I want an M&M!"

"Diep, go eat an M&M. Your mom is literally in Vietnam. She can't
force you to go to a personal trainer if you don't want to. What's she
going to do? Fly here to make you do push ups? Ignore your mom and
live your life."

"Alright," Diep inhaled a clarifying breath to stop the crying.

"Alright," I hung up the phone. I stared up at my boyfriend's bedroom ceiling and decided that I would stay friends with her until she got better. I had to see her change. I knew she could, I would just ride it out, I resolved.


I needed to see Diep recover because I saw myself in her.

The combination of my parents' divorce, the consequences of my early puberty, and the dangers of technology drove me to a serious psychological break when I was twelve. At least, these changes all occurred at the time I unhinged. Because my parents had so lovingly tended to me when I had cancer as a child, I decided that I would fake a serious medical condition to get them to come back and be a family again. I saw a seizure on Criminal Minds and decided I could move my body like that. I faked a seizure at school.

Unfortunately, I'm a brilliant actor, and I faked those seizures very well. The school called an ambulance. My parents came running. The doctors believed me. I didn't have to go to school, which is great because I've always hated school. The doctors said, "We don't know what's wrong; we'll do more tests!" and my parents were nervous and united again by my side. They ate dinner together. My dad came zooming across Metro Detroit any time I was sick. From my perspective, I had found a brilliant solution to the problem of my parents separating.

We could go back to being a family.

I could go back to being a kid.

So, I faked a couple more seizures. No adult in my life stopped me. Not the doctors. Not my parents, who trusted the doctors. Not anyone. Surely, the doctors knew I was faking them, right? Surely, their negative tests on every possible cause of seizures would have indicated I was lying, right? I didn't know that people were supposed to lose consciousness when having seizures, so I stayed awake talking the whole time, and the doctors knew that aspect of my case profile, so that would have been a good indication that I was lying, right? But the doctors who treated me led my parents on. They told my parents that the medical condition was real when it wasn't.

It hurt my family. It made me unpopular in middle school. It hadn't occurred to my twelve-year-old brain that my reputation might suffer from my plan. I couldn't go to regular classes and so I had to admit to myself that I was a liar; that this madness wasn't going to stop until I stopped it. A competent doctor would have known I was faking those seizures, but they led my parents on like there was something medically wrong with me until I fell so far into my lie that I had to resuscitate myself.

Like all fallible people, I am most responsible for my decisions and my lies, and the greatest source of my shame is the fact that I did it. My decisions. My mind. However, my doctors were negligent. The doctors could have saved my social life for the rest of my adolescence if, the first time, they just told my parents I was lying. They could have saved my relationship with my family—and my family forgave me, obviously, because that's what families do, but the doctors could have saved it before it got bad. I have felt terrible, terrible shame for too long. I have been drowning in a guilt so thick I couldn't bring myself to say the word 'seizures' out loud for the past decade. I couldn't say the word 'seizures' until I mustered the courage to write this story. That's what Diep's doctors were doing to her. They were negligent. They were dishonest. Diep's doctors knew very well that Diep wasn't suicidal—everyone who met her saw how she prioritized herself and what she wanted and her, her, her every time—but they kept treating her for suicide anyway, insulting the other patients. However, Diep couldn't connect that her self-hatred was actually shame, so she couldn't resuscitate.

I needed to be Diep's friend because I needed to see her overcome this. She was sick, but not with what she was getting treated for. I would be the person who stops it, I resolved. I would help her. Diep and I are both victims of our own lies. Diep needed a friend who really knew that love stops lies.

A couple days after Diep's mom fat shamed her, her roommate found her foaming at the mouth in the middle of their living room. Ryan broke up with Diep because he didn't want a relapse of his own mental state. I stayed Diep's friend at fifth "suicide," which wasn't suicide. I remained her friend.

Diep had bought me a book titled *How to Date Men When You Hate Men*: a guide for feminists. I didn't hate men and I thought the book was stupid, but I said, "Hehe, yeah, I love it, thank you!" That was around the time I was getting really serious with the man I intended to marry. Diep never considered such commitment for herself. As she continued to hook up and make many more irresponsible decisions, using apps that should not allow unhealthy, self-harming behaviors but do, I texted her a picture of an aborted fetus. I said, "You know that this is real?" It showed a picture of a little red baby with eyes, nose, a rib cage, fingers, toes, guts spilling out onto the surgical table. It was in the first trimester so its head was disproportionately large, but it did not look like yeast in a test tube like I had imagined abortion up until I looked at a picture of a little red dead baby.

Diep texted, "LMAO looks like far-right propaganda."

"No, that's a real picture," I replied, "I feel betrayed that no one ever made me look at that before. You know, so I could understand what I was saying when I claimed that I was pro-choice. The medical establishment would have, through an appeal to authority and the use of jargon, convinced me that I had not just killed my own child. If you kept acting like this long enough, eventually it would happen. You must confront the reality of what you're doing."

"You need to keep politics out of our friendship," Diep warned. "I know people who've been raped." You mean like me, bitch?

Diep was not pregnant. She would have told me if she was.

If you love someone, you will not let them hookup unprotected over and over again. Hooking up unprotected will inevitably, predictably result in either a child being born in unfair circumstances, or a child being terminated. Both of those choices are wrong. If you have the choice to save a child from poverty, that's the right choice. If you have the choice to prevent a baby from being decapitated in their mother's womb, that's the right choice. The fact that Diep could fall back on r/childfree quotes to justify what she was doing as correct is as bad as a person shooting a gun in a public park and quoting Second Amendment rights. How dare she. So I decided this is it. I said, "You need to confront reality. You need to stop hooking up with people. Your actions have consequences, and you are going to have to choose to kill another person. You need to stop."

Diep said the friendship was over. We blocked each other. I haven't spoken to her since. After I stayed with her for four years, Diep threw me out like a piece of trash over this confrontation.

I anticipate that some people will accuse me of not being properly feminist—that I got what I deserved—because they love abortions. I once got into a fight on Twitter with a woman who said she would get pregnant just to get an abortion and send me a picture of it. I posted the same photo I sent Diep on the thread. She looked at the baby, with eyes, nose, a rib cage, fingers, toes, guts spilling out onto the surgical table, and said she would suck it like a gummy bear. She wondered if its bones tasted like pretzels. Then I got suspended from Twitter, not her, for having posted the photo. Twitter didn't care that she would crunch the bones of a fetus like pretzels. The moderators didn't think that violated community guidelines. This interaction happened during the 2022 primaries, and this baby-eating-advocate really believed that she was morally right, just like Diep felt morally right. They feel right because they are insane, dark creatures whose souls slip and flip and blow like the nothingness they are, and I do not apologize for offending them.

Despite graduating with an abysmal GPA, Diep has been accepted into one of the most prestigious art programs in the country. She lives in California and is now connected to big publishers. I can't help but think that's why people hate elite school graduates. These elite programs keep accepting rich kids like Diep who clearly didn't deserve to get in—who don't know up from down, but they tout them as the rising leaders? Does anyone want to be led by Diep? Who knows, maybe Diep will run for governor of a state one day.

Even though I saw myself in her, we weren't the same. I was twelve when I faked those seizures. Diep was twenty-two when she faked her suicides. I was a child. She was an adult. My motivations for doing what I did were methodical, based on experience, and goal-driven: I wanted my parents to get back together like they were when I was sick, so I faked a sickness. Diep faked suicides to emotionally dry heave, occasionally to test how much everyone else cared about her. Diep is not that smart. But Diep was enough like me that I saw her for who she was. I desperately tried to save her so that I could save myself from my memory. If I had been successful over the four years I tried to be her friend, then I could have contextualized my seizures into the story, "God gave me that soul sickness so that I could help another person with the same soul sickness." But Diep wouldn't fucking listen to me and she fucked up my healing process.

And that's why faking suicide is a powerful tool for manipulation and revenge. Diep's mom probably wanted to think that she was a good mom. Diep hated her mom. Diep began faking suicides and Diep's mom drowned in guilt and worry, and it tasted like sweet revenge. Grow the fuck up, Diep. Your depression is hurting the people around you. You're doing it because you know that you're hurting us. You want to hurt us. We see it for what it is.

We see you for who you are.

Just stop it.

Just stop.

So, our relationship just ended. Poof, gone. After all that, *she* cut *me* off. I can't give you a better resolution to this story because she didn't give me one. I would like to tell myself that God put Diep in my life to shape my worldview so that I could help those who haven't yet confronted themselves for who they are, yet. Perhaps God wanted me to condemn contemporary American culture's bizarre reliance on doctors to heal our souls. Mental health care in the United States sucks—not because there isn't universal access to it—but because it is ineffective. If people could go to their doctors for mental health solutions, then the rates of depression and anxiety would be lower. If people could go to their doctors for mental health solutions, then there would be solutions, not symptom mitigation tactics like long-term reliance on therapists and antidepressants. If psychology as a field was a good way of framing internal pain, then rates of mental health problems would be lower than they are. But the rates of mental health problems are not low. The rates of success in therapy are not high. We rely on so-called experts of the scientific method to heal our souls, but perhaps souls cannot be healed that way.

Before modern psychology, people experienced internal pain just like us, but they used the vocabulary available to them to express it—mostly religious vocabulary. Before people had 'emotions' they had 'souls.' Many non-Western cultures still use the word 'soul,' particularly the Middle East. My Arab grandma, who has maintained a staunchly religious framework to contextualize her internal experiences her entire life, and who is touching something very real when she prays, would never use a word like "depression," but instead words like 'guilt,' 'shame,' 'grief,' and 'sorrow.'

What is the difference between 'depression' and religious words like 'guilt,' 'shame,' 'grief,' and 'sorrow?" The difference is their implied cause. The implied cause of guilt is that my grandma has somehow disrupted her relationship with God, and she needs to make it right. The implied cause of shame is that she has somehow disrupted her relationship with her community. The implied cause of grief is that something unjust affected her. The implied cause of sorrow is more existential, so it is more like 'depression' than any of the other words in her arsenal. But the direction of 'sorrow' is very much pointed up. When she's talking about sorrow, she's talking to God.

Because my grandma—the one who bought and made the vegetarian turkey for Diep, who failed to come to her house—has ordered her internal life around a religious framework, and uses a much more precise vocabulary to describe her pain, she almost never has to sit around wondering why she feels sad. Diep pontificated, "Why do I hate myself?" all the time. Hell, I pontificated, "Why do I hate myself?" all the time, until I became religious and ordered my insides correctly. The words my grandma uses prescribes the cause and sets her on a path towards solution. If my grandma feels guilt, she apologizes, and it goes away. If my grandma feels shame, she makes peace, and it goes away. But many internet people who suffer from depression float in their egos, asking, "What is happening? Why am I depressed?" utterly sideswiped as to what might have got them there.

I do not believe in conspiracies, but it seems to me that if I wanted to control a population, or sell a lot of antidepressants, I would strongly prefer that the citizenry use the word, "depressed," rather than 'guilt,' 'shame,' 'grief,' or 'sorrow.' If the rabble use the word 'depressed,' they're confused, and leaders could exploit their confusion. I, the King, could theoretically say to my citizens, "The reason you're sad is because of our racist institutions, so help me destroy these institutions," and then while they're trying to destroy what I tell them makes them sad, I can build up a new system in my image. I am not saying that I think our government has somehow purposely engineered this language shift. It is simply the fact that if, hypothetically, I was a maniacal dictator, I would be very happy that people stay floating in the vague and heavy state called 'depression.'

My Arab grandma—the one who I stayed with while I was processing Taras—says that pride is not a solution for feelings of shame. She uses the word 'pride' to describe 'when you put yourself above others.' In this context she uses the word shame to describe 'when you can't stop thinking about how bad you are,' or, 'questioning if other people like you.' My grandma says that pride will not solve shame. Effectively, my grandma is talking about self-esteem and depression. In translation to secular language, my grandma was trying to tell me that self-esteem is not a solution for depression. It will not work. Putting yourself above others will not make you love yourself more. Diep's therapists specifically instructed her to prioritize herself above her friends, as if Diep could dilute self-destructive behaviors with self-centered behaviors, kind of like how you dilute acid with water. Unfortunately, the human soul cannot fit into a test tube.

If Diep was a paper doll, my grandma would be a ball of lead. If Diep was a plant, my grandma would be a garden with orchids and lavender and fig trees. My grandma is just stronger and healthier and deeper and better. If she behaves wrong, it puts her wrong with God, so she has to go back and fix it. But at the end of supplication and apology is redemption. There is no similar secular version of that process, besides apology, which people don't put much stock in, anyway. Diep and I never thought of ethics as having anything to do with self-esteem, but my grandma cannot separate those ideas because ethics is how she accesses herself. Ethics orient her self-esteem. Religion is, for her, the practice of ordering her insides with ethics, and it has made her insides strong. With an ordered inner-self, she is a wellspring of love. This is not a religious metaphor. It's an anthropological observation.

In that same line, I look at Diep, at me, at our peers, and I think that if the field of psychology was so effective at healing people, then in an age where the field of psychology is dominant, then the rates of mental health problems would be low—but they are not low. If psychology was a productive framework for ordering your internal life, then more people our age would have an internal life like my grandma's—but they don't. If the field of psychology was so effective at healing people, then the team of maybe fifty so-called mental health professionals who saw Diep's file over the course of her total medical treatment would have been able to find a cure—but they didn't. When are we going to admit that putting soul-care under the umbrella of medicine is a mistake?

# A MEDITATION ON FREEDOM

On Google, as of September 26, 2022, freedom is defined as "the ability to do what you want without restraint."

## PART 1:

Eight billion is a very large number. If you stood in a room counting out loud but never slept or rested, assuming it takes two seconds to pronounce each number – one, two, three, four, and so on – then you would reach eight billion in 507 years.[1] That is how many people live on this planet.

All eight billion of us know people we love and have memories we hate. All of us want to be remembered, and all of us will be forgotten. You're one person on one planet in one solar system, and there are trillions more solar systems. By an argument of volume, you are so infinitesimally small that you will not—cannot—matter to anything material. No person's grave will last. No sculpture on earth will last. The inhabitable earth will not last. No one person's successes or failures or pride or problems can mean anything on the scale of what exists. You are infinitesimally small, and the world will get on just fine without you.

## PART 2:

When you stand still, you're moving through time, which can be mathematically treated the same as moving through space. This is the brilliance of Einstein's theory of relativity: the spacetime continuum—that

---

1. 8,000,000,000/ (30 words per minute* 525600 minutes per year)

time and space are the same type of dimension.

As we have already established how small you are in relation to the material universe, so, too, we establish how unimportant your life is on the scale of time's eternity. If you're a layperson like me, your whole memory will probably be erased within four generations. If this book remains for two hundred years, then I might live a little longer through my writing. But eventually, war will break out or this book will become obsolete. Maybe it just won't sell well. Then, I and my memory will be lost forever.

So too will the memories of Alexander the Great and Barack Obama and whoever else who seems famous and everlasting. Just as we don't know the biggest, baddest dinosaurs of the Triassic era, in a couple of million years all the stories of the age of man will be forgotten. There will come a time when every grave is erased, every mountain moved, every living creature gone because the sun turned into a ball of iron and stopped burning. Even if you carve your face into stone, eventually, the stone will erode. Human civilization will end, and all our leaders and problems and certainly all the petty shit you deal with will eventually die. You will never be remembered on this earth. No one will.

Time is longer than our memories.

## PART 3:

So you are small and mortal. Ironically, one could say that being small and mortal is an eternal and infinite fact. The sooner you accept that unavoidable fact, the sooner you can occupy your mind with purpose and meaning to life. But let's stick with you, the small and mortal, for a moment, in the context of civilization.

In many tyrannical regimes throughout history, you had to obey the regime or it would kill you. Therefore, in order to become free, you had to overcome the fear of death. Christian martyrs and Japanese samurai alike created a culture about not fearing death in order to maintain their autonomy and freedom. And many of them prematurely walked to their own graves singing.

In the twenty-first century, Western regimes do not threaten death very often. Unless you are like Jeffrey Epstein, who threatened the reputations of the powerful, the order that exists doesn't need to kill you. If you are a layperson, mostly controllable anyway, all a power needs to do is threaten to kill your legacy. If you're convinced that your legacy will be tainted unless you obey the standard order of things, then you will obey the standard order of things. If you believe that you will live in eternal shame unless you adopt a certain set of beliefs, then you will adopt that set of beliefs. That is why religion (or the political beliefs that substitute for religion) claim a monopoly on eternal shame. If people feel shame for not conforming to them, then most people will conform, because most people hate shame.

But if you realize that the power of those who claim to control your legacy is finite, as all human institutions are finite because no human institution can have any meaning on the scale of space or time, then you can start to grasp the smallness of your own leaders. Without perspective, your leaders can seem like the whole universe. But they're not. When you realize they're not that important, you don't have to obey them. This is why realizing your own infinitesimal smallness is the first step in becoming free. Accepting shame, like accepting death, threatens order and power.

Governments use death for power.

Now, shame substitutes for death.

Governments (and religions, and the political beliefs which substitute for religion) use shame for power.

## PART 4:

In communist Bohemia around 1960, calling someone an 'intellectual' was a terrible slur. It meant that you thought yourself above the people, and you hanged for it, said Milan Kundera. In a way, Kundera observes that dying with your feet off the ground memorializes the crime of floating above others.

Kundera was not hanged. But he wrote something critical of The

Party and was rejected by his own country. First, they said he couldn't write. Then he couldn't speak in public. Then people who wrote with him and spoke to him couldn't write or speak either. Then he ruined a young woman's life by asking her to publish something he wrote under a different name and it was found. He left his country. Outside of communist Bohemia, sliced in half with grief and loneliness, Kundera could now write anything he wanted. His country's rejection of him gave him total freedom. So he freely wrote about grief and produced the two best books I have ever read: *The Book of Laughter and Forgetting* and *The Unbearable Lightness of Being.*

In *The Book of Laughter and Forgetting*, Kundera contextualizes the pain of exile from his home with the metaphor of a ring dance. Kundera notices that a circle is a closed formation, and that when you dance in a circle every happy participant looks into its center. If you leave a row, you can come back into it from either side. But when you are cast out from a circle, you cannot come back into it. It is no coincidence, Kundera says, that our universe works in circles. "When a body of matter breaks off from orbit, it will float inexorably away, carried by centrifugal force; and when someone is cast out of a ring dance, there is no getting back in."[2] You are either in a formation or out of it. The formation will be your home, but the space will be your freedom.

Using his metaphor, I will assess that I was too clunky to ever step to the beat. But being cast out actually set me free. An example of freedom by rejection is my career. Up until the beginning of COVID, my senior year of college, I had assumed that I would get a Ph.D. in writing. But I was not successful enough at connecting with my professors to get real letters of recommendation. I did not navigate the social milieu of academia to pull the strings in my favor. I did not do as well on my thesis as I would have hoped. A good mark on your undergraduate thesis is a necessary prerequisite for most competitive Ph.D. programs because you have to show that you are a competent and disciplined researcher. I didn't know how to produce what my advisors wanted. I got mediocre

2. Kundera, Milan. The Book of Laughter and Forgetting.

marks on the thing that was supposed to propel me into my career, and I didn't have other research to back me up, just a stupid list of clubs. Without the thesis, I withdrew my Ph.D. applications. For about a year after, during the height of the pandemic when we all covered our faces, every time I thought about my thesis I would sob behind my mask in my room or in the Publix holding a box of flax seed crackers. One time during the end of quarantine, while I was walking my chihuahua in the drain tunnel woods, the pattern in the bark of a tree branch reminded me of an Arabic letter, and I thought of my failed thesis. My legs gave out under me. Belly down, my left cheek in the mud, I laid still in the dirt for over an hour as I resigned to my lot in life: I was an insignificant loser. I lost. I was out of the game. I had no connections in publishing. I wouldn't get a Ph.D. I had no way to be respected by those institutions which writers are supposed to operate in. I had no connections from AWP. I would never get a chance to show my class that I was of the same stuff as them.

This was when other friends and I broke up. Of course, the only common factor in all of my failed relationships was me. I had gone into the college with the childish expectation that I needed friends, and I surrounded myself with people who were not actually capable of soul sync because I wanted the general social milieu to perceive that I had friends and status. When these people and I ended up not having anything to talk about, or I felt disgusted at their utter stupidity, I would get bitchy. Some of the girls, I blocked. Other girls blocked me in retaliation. More still stayed in contact and then eventually dropped off. More than anything else, my loneliness was my fault. Yes, I could have been a better friend to these people. Yes, I was very angry when I knew them. More importantly, I could have released myself from the desire to appear like I had social status. Being part of a group I had changed myself to fit into did not alleviate my loneliness. On my 22nd birthday, I had a great Zoom birthday party with 54 people, all of whom I at the time considered my close friends. At the age of 24, I still speak to three. This is because it is not friendship if a group accepts you for who you're pretending to be.

Unsuccessful. Alone. I would go get a shitty little job and live my shitty little life, friendless, second-rate, award-losing writer with no discernable path forward. So I kept laying in the dirt. I nearly fell asleep, and began dreaming in voices. Eventually my dog licked my face enough that I rolled over, rubbed my eyes. That night I wrote the flashback scene in *My Brother the Fanatic*, completing it. My pain made my art. My suffering served a higher purpose because it produced something beautiful. Awards aside, I could still tell a good story.

Two years removed, I can now reflect that if I had gone into Academia, I would have been taught to be a critic of high literature—produce a Freudian analysis of Lope de Vega, apply a Marxist framework to *The Yacoubian Building*, fight with other academics about whose interpretation is best, criticize and tear apart whatever I'm reading—but I would not have been taught how to create. I would not get to choose how I wanted to write or who I wanted to read. I would not have been paid well. I would not have been allowed to express any criticism of race and gender theory even though there's a lot to criticize. I would have been subject to the whims of my academic advisor; and if that advisor happened to be a dictatorial bitch I wouldn't have a lot of protection from her. As I distanced myself from the institution which I thought would be my home, the editor's voice in my head began to shut up, and I developed the ability to write without shame.

So, what did I get from being rejected by what I thought was my future?

Outside of the ring dance, I found creative freedom.

## PART 5:

In the twenty-first century, social media makes us feel infinite. The Internet is infinite, so we think that if we put ourselves on the Internet then we, too, will be infinite. While we are wrapped up in our own egos, cultivating reputations online, we fear losing that reputation because it will feel like death. In a way, losing your reputation *is* a form of death: a loss of self, an ego death, the letting go of your control over your own legacy. Ego death is not easy for me, but it started happening in my twenties because I was socially rejected. It was only after the ego death I began to develop social awareness.

I had a solid group of six girls in high school, but that broke up when we got to college and I didn't make another cohesive group like that. I fought with my roommate. I joined a sorority that neither fulfilled me internally nor gave me the social status I thought it would deliver. My colleagues ostracized me and they bullied my coworkers. All three of my so-called best friends at separate times broke up with me for different reasons. If that wasn't enough, how else do I know that I was unpopular? I can quantify it. On Instagram, I pulled between 80-130 likes per post in my prime, sometimes as low as 50, while most girls in my social circle pulled 200-400 likes per post. I was one-third as popular as my peers on average. Night after night, I imagined standing on a stage, watching a line of attractive men pick 70 other girls before me in a room of 100. I saw a sexy blonde man with a chiseled jaw take a skinny girl with a long face and I felt so bad about myself that I started thinking terrible, vapid things about other girls on Instagram. I was jealous of every accomplishment they ever posted. When I was 22, I messaged my friends to get my like count up on my graduation post. Isn't that sad?

Then I decided *enough of this shit*. I deleted my social media. I finally turned off the incessant echo chamber filled with other people's political opinions, especially privileged people—and everyone who graduates from Emory is definitionally privileged regardless of how they grew up—pretending like they have a clue about hardship, selflessness, and sacrifice. When I was on Instagram, I was part of their discourse. But when I got off Instagram during COVID, I started working at the Amazon warehouse, and I compared the people I met there to the "underprivileged" people at Emory. A lot of the people I knew from Emory posted about how they were very underprivileged, unlike the kids whose parents paid for their college, because they were on scholarship. Here's the problem: if you're successful because of a scholarship, then you're not self-made. Your college made you. Whoever decided to give you money made you. You're not incurring the debt yourself. If you are on a scholarship, then you have the backing of wealthy institutions and therefore the same economic status as any kid whose dad works at KPMG. If you came from nothing and still got to the top, then society has not screwed you over. If there are 10,000 diversity and inclusion officers ready to fight for you the moment you experience a microaggression, then you're not in a racist institution. Universities are the most grandiose displays of our nation's wealth, and if you have been successful in them then you are not an underdog. You need to stop lying. You are not capable of leading revolutions against the institutions that built you.[3]

On Instagram, ineffably privileged little shits competed about who is the biggest victim. At the Amazon warehouse, people didn't want to be a victim because they had real experiences with hardship and sacrifice, and people who actually have been victims hate victimhood. Privileged people feel allured by victimhood. Victims don't. And you know which politician most of the people I knew at the Amazon Warehouse preferred? Not Biden. Young adults with elite degrees and scholarships have not been screwed over by society. Young adults throwing boxes onto a conveyor belt in the back of a truck have been screwed over by society, and those young adults throwing boxes overwhelmingly voted for Trump. Square that circle. Rich people think that Democrats are compassionate to the least of these. Poor people know better. If impoverished people don't feel represented by your politics, then your politics don't represent impoverished people.

The process went: I was not popular, it hurt me, I decided to let it go, I had new experiences, and suddenly I started thinking for myself. Freedom by rejection. When I got off Instagram, I could start asking questions like, "Is family important?" "Is religion worth keeping?" "Should I build a marriage with my best friend, like my grandma did?" All through high school and college, I had been told that the reason there

3. It's good to be successful. The problem is not that people are successful. The problem is that college professors and reddit warriors alike are participating in a nationwide delusion.

is injustice in American society is because boo hick Republicans will not get on board with progressive social reform. If they would just let us ban guns, there would be no more gun violence, for example. It was the racists that spurred violence, corruption, and injustice; but if they would just submit to our social movements then those problems would disappear. Therefore, if we couldn't make them submit, we would at least shut them up. But I lived in liberal cities, public schools, liberal universities—places Republicans had long since stopped speaking—and yet our culture was still riddled with corruption and injustice. So I thought to myself, "Hmm, is the reason Democrats haven't delivered a utopia yet because 1. an impoverished and irrelevant person said something racist, or 2. there's actually something wrong with the capabilities and beliefs of Leftist leadership?" I decided it was the second. I disassociated from Democrats, which had been my faith and the faith of my people.

If I had had Instagram, they would have shamed me straight.

## PART 6:

So, social media controlled my political thoughts and kept me attached to the desire to control my own image. Because I was unsuccessful on social media, I deleted it, my political thoughts became rooted in real observations instead of conformism, and I was able to shut up the editors voice in my head and just write. Some people might not like what I have to say, but I'm saying what I really think. There is a direct connection between what I believe and what I'm saying, and that line is authentic use of my energy. I broke away from the ideological reigns I grew up with, and found freedom in thinking for myself. That is good and authentic action, and therefore I centered my energy.

Energy can neither be created nor destroyed. Energy is infinite and eternal. The energy within you is infinite and eternal. It will never go away. It will never cease. It will never die. Your access point to the infinite and eternal rests in your energy. When you were conceived, energy entered you. When you die, it will leave you. But the energy within you right now will last forever. And, once you really grasp that, you come to realize that you are that energy. My grandma likes to say that she is not a human having a spiritual experience, but a spirit having a human experience. That's basically the same idea. You are your energy. You are your spirit.

I had said that 'you are small and mortal' is an eternal and infinite fact.

Therefore, eternal and infinite exist.

Specifically, they exist in energy.

## PART 7:

Your energy is eternal, so how you use your energy is how you engage with eternity. You spend energy when you take action. The secret of engaging with the infinite rests in how you choose to act. The way you choose to act is the only way to engage, to goad, to tickle, to poke eternity. The transcendental purpose of any life is to make choices about how to spend energy.

Not all actions are the same. A mother taking care of her sick child seems to be in one category of action while a mother prostituting her child for drug money is in another category of action. We simply cannot shake off the feeling that certain actions hold attractive or repulsive qualities. Though we may not always agree on what actions deserve what level of praise or repudiation, some form of acceptable and unacceptable, good and evil, has permeated every culture, people, and time. "There are only two things that fill me with wonder: the stars above and the moral sense within," said Kant.

How permanent is the moral sense within? How real is it? I will paraphrase C.S. Lewis to make a point. When someone tries to trip you, you feel wronged. If you're a normal person, you know that you have been transgressed against when someone attempts to harm you. It is not the fact that if everyone went around tripping people the world wouldn't function, a utilitarian argument. It's not the fact that you were raised to be hurt when people trip you, an absolutely relativistic view of culture. If someone accidentally tripped you, they wouldn't have actually transgressed against you. It's an accident. But if someone intentionally tried to trip you, even if you did not fall, they would have transgressed against you. They tried to hurt you. Someone trying to hurt you, even if they don't succeed, is much worse than someone accidentally hurting you. Why? Because morality exists a priori. It never lives or dies. And if you disagree, then you have no ground to say that you were ever mistreated.

We engage with the eternal by spending our energy, we spend energy when we take action, and actions can be mapped on a scale of right and wrong. Humans have a specific understanding of right and wrong. We can choose whether we want to adhere to or ignore the rightness of our actions. We can freely choose to do good. I can choose to direct my energy towards moral action or not.

Specifically, morality is a direction energy can flow.

## PART 8:

By approximately 4,000 years of accumulated human knowledge, since the beginning of the Abrahamic faiths, people have been telling their children that their immortal soul depends on following the commandments. Perhaps this idea is not dumb. It is at least worth considering merely because of its vast historical precedent.

When I followed the religious commandments as my grandma explained them to me, within five months—maybe as little as eight weeks, but I fell out and came back—I developed a real peace and joy inside of me that felt like living water. I got up and prayed. I listened when people were trying to tell me something. I supported my mom. I let go of minor irritation. I forced myself to say a prayer for anyone my memories wanted me to hate. I asked God to help me remember that the person who cut me off in traffic was a person with a family, not a crustacean who exists to irritate me. I made a weird girl feel accepted. I told a popular girl who would have previously intimidated me that she was beautiful and meant it. I felt guilty about mistakes I made three years ago, then I imagined picking up a toddler version of myself and saying, "It's okay, you'll do better next time. You have to learn in order to know." When I evaluated myself based on my compliance with religious commandments, I never had to wonder if I was good enough. I was as good as I followed the commandments.

With religion, I never needed to pontificate about my self-esteem, so I stopped falling into cycles of self-loathing. I stopped evaluating my career based on worldly successes, and I connected to immeasurable joy in evaluating whether or not I was a force for good. With religion, I always knew how good I was or wasn't. If I hadn't been good, I had to admit I was sorry and make it right, but at the end of that tunnel there was redemption and forgiveness. The thing that made me feel the worst about myself was not how others mistreated me. I had low self-esteem because I accurately recognized my own fallibility, stupidity, and sin. Then I found God, and that fallibility and sin, when recognized, brought me closer to Him. Even my worst mistakes could be used to center my energy. About five months in, my internal life felt ordered, deep, and connected.

So perhaps our ancestors said that our immortal souls depend on following the commandments because they knew something about immortal souls. Perhaps they were imparting the most brilliant, robust model of existential purpose to have ever existed.

Perhaps religion is important.

## PART 9:

As opposed to a monarchy, which expects blind obedience to a king (who could order you to compromise your soul, and you would have to do it or die), the American founding fathers would design a system of governance which respects your right to your soul. The government does not need to be our dad. The government will not make a person recant their religious beliefs or utter falsehood. We, as autonomous, thinking individuals capable of compassion and rational discernment, will regulate our own behavior according to transcendental morality, and our government will not infringe upon that. That is what they meant by freedom.

The American founding fathers spent a very long time designing the checks and balances which would prevent a king from taking power. But the idea that the society for which they created this freedom-respecting government would lose the concept of the soul? That 250 years later people would be thinking that 'freedom' only means your ability to eat cake pops and binge-watch *Better Call Saul* if you want to? That 'freedom' means to limitlessly indulge yourself? That all their work put into 'a place where people don't have to compromise their souls for the government' turned into 'a place where people can do what they want without restraint?'

That would have been unthinkable.

## PART 10:

Now we get back to you, the small and insignificant person, like me, reading this book. If you want to be free, you must do three things:

1. Release yourself from the fear of death, or the fear of ego death

2. Cultivate morality in your actions

3. Notice the soul that begins to develop

I tried it and I think it's a model that works.

In conclusion, screw Google's stupid definition. Real freedom is not hedonism. It is not any form of addiction or indulgence or self-flattery. It is not your ability to do what you want without restraint. Real freedom is a system of governance which does not ask you to compromise your soul.

# ON ACCEPTING INJUSTICE WITH AGE

January 2015
Dearborn, Michigan

A month and five days after my Sitto passed on, my nephew Isaac turned two. He didn't understand his birthday party, but he loved fitting his new plywood animal figurines onto the green puzzle board. Lying on his stomach, my sister's baby turned the alligator block around sideways so that it just rimmed the space it fit into. My mom said that Isaac could concentrate better than any baby she knew.

My sister Alayna moved to help Isaac fit the piece, but our mom chimed in. "No, he's figuring it out!" We watched closely for a long time until Isaac finally popped the alligator shape into its place on the board. We cheered for our baby.

"Maybe Isaac can be an engineer one day, like Baba," my mom boasted, referring to her own dad, not Alayna's or mine. Sitto's husband.

"Isaac'll be whatever he wants," Alayna corrected, staring at her son. Our mom stared at Alayna with her eyelids half down, contemptuously. There was no reason for Alayna to imply that our mom was predestining his career. Alayna was just condescending sometimes.

The gears from the garage growled as my Uncle Elias, my mom's brother, walked in through the back door. He noticed the block in its proper spot on the puzzle board. "Hold up, little man, you did that?!"

Isaac smiled.

"That's wild, man!" My Uncle leaned in for a fist bump and ended up just holding his great nephew's chubby hands. "You're getting so old, little man." Uncle Elias played with the tips of his baby fingers. "When are you gonna start helping with rent?" He asked.

"Yeah, Isaac," Alayna agreed. "Why don't you have a job yet, boy?"

Alayna didn't change her tone between sarcastic or serious, so Isaac babbled happily. Nonetheless, my uncle mediated, "Hey, now. How 'bout this, man? How 'bout we pay for you right now, and then you get a job when you're older?" Because Isaac was a baby, my uncle answered his proposition himself. "*Akeed.*"

My mother took Isaac from her brother and rubbed her lips in his curly hair.

Some weeks after that, in United States History class, my sister texted me. "I'm picking you up, we're going to Canada." The legal drinking age in Ontario is 19 years old, my sister's age, so we had gone a couple times to drink. While everyone else was scrambling to their next class, I slipped out the back door to the pickup zone. As I climbed into Alayna's tan Camry, I asked, "Who pissed you off?"

"School."

I didn't push the question further.

Alayna had switched to a community college GED program after she got pregnant. Three years older than me, she had already earned her GED and was pursuing her Associate's at the community college. She planned on transferring to the University of Michigan Dearborn next year, majoring in anatomy, and then going to medical school. Alayna was an incredible student, much better than me. Alayna's got the memory of God. I'm not kidding. I could say, "What's that song where we danced like Egyptians?" Then she'd rattle off, "I could really use a dream or a genie or a wish to go back to a place much simpler than this . . ." And she could do the same thing with textbooks she'd read.

Before COVID, it was very easy to cross the border into Canada. We waved 'hi' to the Canadian guard wearing a park ranger hat and drove to the closest tavern in Windsor. It was a building made of dark-green wooden planks and three small, high windows, ready for the high snow. Alayna ordered us beer. Alayna sat there silently for a long time, so eventually, I tried to break the ice. "Do you like the new Curly Girl conditioner mom bought?"

Alayna's lower lip trembled, so I grabbed her hand. I tried, "Who do I need to beat up for you? Is it Charles?" I asked, referring to Isaac's dad.

"No," Alayna replied, which meant he wasn't good but not particularly causing problems right now. Alayna tried to talk, but a jolt in her throat stopped her. It looked like she was swallowing marbles. We just sat in silence sipping beer.

Finally, Alayna told me, "I'm gonna fail in Spanish."

"I thought you've gotten As on like every test."

"Yeah, I did. That's what I tried to tell her when I talked to her. But five absences means you fail."

"But two of those were because you were taking Isaac to the doctor?"

"Yeah."

"Well, aren't those excused?"

"Doesn't work like that in college. They don't have to pass you."

"One class isn't going to stop you."

"No!" Alayna cried with an ache in her big black eyes. "No, I can't. I tried to explain it to my Spanish teacher about Isaac, and she said 'Look, policies are policies,' so I went to the advisor to talk about it, and the advisor said, 'If you're too busy with your son to get to class, a place like the University of Michigan doesn't want you anyway.'"

I sat nodding. However pejoratively phrased, I silently thought to myself that the advisor did sort of have a point—if you don't meet the requirements, you don't get the job. But then I shut that down, remembering it was my job to help my sister. Family first. I raised my voice a little. "Again, so what? Prove them wrong! *Ma fe mushkila*. Sitto always said if you fail, learn from the failure." I could see my sister looking down, lips curling and trembling. "Oh, come on Alayna, who cares? School is whatever." I continued. "This isn't going to stop you from getting your degree. And if you don't get a degree, there are a lot of people who have high paying jobs without degrees. When a door shuts, a window opens, right?" I scooched over to her side of the booth and rubbed her back. "No one's mad at you. Mom will understand," I cooed.

Maybe Alayna feared Mom yelling at her. But my sister kept sobbing and drinking and sobbing until I realized I'd have to drive us back across the bridge on my driving permit.

Luckily, border security pushed us through without making us roll the windows down. I thought that Alayna had overreacted, but I guess we all have those days, and I didn't think much of it.

Alayna's dad, a Yemeni named Abu Bakr who looked like an Indian, left my mom for another woman back in 1997 when Alayna was still in the womb. Pregnant with Alayna, our mom moved in with Sitto and Jiddo. Jiddo died in 2001. With the patriarch of our family gone, and one failed marriage, our mom began dating like an American, and she found my dad, Jackson Galloway, the whitest man this side of Kentucky. My sister came out slender and tan with a low, wide pelvic triangle. I came out splotchy, green-white, long torso, short legs. Photo after photo of my sister and I in childhood showed Princess Jasmine next to Frodo the Hobbit. My sister went through puberty around fourteen years old, and since then men have been asking to buy her things and to take her places. Grown men, like bus drivers – not just the boys in our school. I was jealous for a long time. Then I realized that my sister's beauty cursed her.

We closed the car doors and started off to Ecorse to confront Charles for this month's child support, which he hadn't paid. Charles might be incompetent but he loved his son, so my sister found herself in the awkward position of maintaining a perpetually dysfunctional and codependent relationship with someone whom she could not leave. Charles would have been great if he had married my sister and lived in the same house as his child. Charles would have been good if he didn't quit his job every two months. But Charles was neither great nor good. Just a person who impregnated Alayna two years ago, but who, unlike Alayna's dad, loved the baby which their sex produced.

We pulled up to the Ecorse house. Alayna clambered out of her tan Camry and walked up to the front porch. I stayed in the passenger seat with the windows rolled down, phone in hand. I knew yelling was normal. But if they started to scream or threaten, I would call the police. Alayna banged on the front door and yelled, "YOU WERE SUPPOSED TO GET IT TO ME LAST WEDNESDAY! I SWEAR TO GOD YOU WILL PAY INTEREST ON WHAT YOU DON'T PAY!" Then she switched to addressing Charles' mother who apparently was within earshot. "Mrs. Williams, your grandson needs his father to act like his father!!"

"Bitch, I am being a father to my son!" Charles stepped out onto the front porch and closed the door quickly behind him, signaling to his mother he didn't need her to intervene. Here we go again. Charles shouted, "Just last week did I not come over to teach my son how to play catch? Did I not do that? I did, for over an hour, I'll remind you. . ." and this went on and on.

Alayna said, "You always have an excuse, Charles!"

Charles said, "You always have a complaint, Alayna!"

On and on and on.

Once more it was Charles' turn. "Good God, you brought your sister here again? That's how little you trust me? I'm disrespected! You don't deserve any of my money! Do you have any idea how rude and inconvenient it is to—"

"I don't care what inconveniences you!" Alayna screeched, face turning crimson. "You want to talk about inconvenience? Because I'm the only one of us who takes Isaac to the doctor, I couldn't make it to my Spanish class!" Alayna started to sob and continued. "You know what the advisor told me? She said 'maybe you should have thought of that before you had him! There's no reason anyone has to carry to term anymore.'. . ." Alayna carried on.

*Maybe you should have thought of that before you had him? There's no reason anyone has to carry to term, anymore.* I died a little inside. So *that* was what had happened the day we drove to Canada. That was why she had been crying. The audacity of that bitch. I imagined my baby nephew Isaac, lifeless on a doctor's table between two forceps prongs. I didn't hear the rest of the fight. A watercolor memory indicates to me that Alayna had walked away with $267.75—as if three extra quarters would make anything better. There was only one thing I could think of on the drive home. *I'm gonna kill that advisor.*

Beyond my conscious control, I imagined pouring boiling water on Alayna's advisor and watching her skin blister and shrivel up to expose muscle, then bone. I wanted that advisor to suffer. I wanted her to scream. I had never experienced blind, consuming hatred—the kind that kills your soul—until that day. Some part of me recoiled at the mental image of the adviser being tortured, but another part of me kept my inner eye fixated on the moment when life would leave her body and the world would be ridden of her forsaken soul. I hated her. She had subjected my sister to the last indignity I would allow my sister to experience, and she rightly deserved to die.

That night, I tucked into bed staring at the picture of my Sitto. I thought of the time my Sitto was disrespected by a grumpy worker at checkout and she told us that when people disrespect us in life, we ignore it because our dignity is not affected by them. *What a load of shit*, I thought. That advisor would get away with what she said if I didn't step in.

That night, I snuck into Isaac and Alayna's room. Alayna was out cold, lying on her back with her boobs spread sideways like frog eyes. Isaac was twitching a little bit. He looked blue in the moonlight. I rummaged through Alayna's backpack and took out her ID. I didn't make noise, but when I stood back up, Isaac's bright black eyes were gleaming at me. I slipped the card into my backpack.

Our mom would demand that Alayna and I treat Charles with respect, even after Alayna and Charles got into screaming fights and hit walls and each other. I should preface this by saying that our mom loved us. When we were kids, she rocked us gently and in the morning she would wake us up slowly so we could remember our dreams. She really listened to what we had to say and made us feel accomplished for the stuff we did, and she took us to museums. She made us both feel beautiful, even though one of us was clearly more beautiful than the other according to worldly standards. For Alayna, whenever she did well on a test, our mom would call Alayna a winner and put it on The Winner's Wall, in our pantry where we pinned up drawings. But then Alayna had Isaac and fought with Charles, and our mom would snap, "*Hade beitee*, and I say he comes in to eat with us."

"He called me a bitch five minutes ago," my sister would protest.

"*Baseeta.*" My mom waved her hand. It means 'simple,' as in 'that's a simple problem.'

"*Baseeta?*" My sister repeated, outraged that my mom minimized her.

"Ey, *baseeta*. You don't need a boyfriend. Your son needs a father. Charles? You in the driveway? Come back in and have dinner," our mom called from the side door.

Alayna hung over my mom's shoulder as Charles shifted his car into drive and pulled forward. "You don't care how he treats me!" Alayna yelled at our mom.

"*Wa la yihimmik. Lazim yansha ma abhu.*" That's tough to translate. It means something between "don't worry about it" and "I accept your apology." Kind of like how moms will say "you're welcome," in a sassy way to remind their kids to say thank you, my mom said, "I accept your apology, your son needs to grow up with his father."

My sister cried, "First of all, it hurts me that you're so indifferent to my pain, and second of all, you're wrong that he's going to be a role model for Isaac. He's one up from a street bum! I want my son to grow up respecting women."

Our mom turned around with her flat hand raised, switching to English. "Isaac did not choose his life. You chose Isaac's life. You owe it to your son to make sure he knows his father." Alayna looked down and nodded, so our mom lowered her hand. "Now get Charles some tea, *hullu.*"

Alayna didn't get tea. She crawled into her bedroom and cried. Later that night, she threatened to move out. Our mom said, "If you do – I'll see you in court. I still have pictures of your *mukhadarat*," she threatened, referring to the methamphetamine she had once found in a dresser drawer. "You're no more fit to be Isaac's parent than Charles."

Alayna denied that the crystals were hers, but they were definitely hers. She wasn't an addict – largely because mom found them early and shamed and humiliated my sister into submission. For what it's worth, it stopped my sister from using hard drugs. And my mom didn't enjoy shaming her like so many other women do. Alayna had tried to delete the picture from mom's phone, but she had made copies and backups. "Alayna, you can go. But my grandson stays. If you try to take him, I'll see you in court."

Alayna and my mom were dysfunctional back then. It was mutual destruction more than love that kept them living together. This always struck me as odd because I remembered when my mom and Sitto would call Alayna the good daughter. "*Mashalla*, so beautiful. So smart." Alayna always got high marks. I remembered as a kid being unable to answer questions, or making mistakes, or forgetting my glasses, or breaking glasses, or having acne – which my sister did not suffer from. Then my sister grew up, and my Sitto passed on, and our roles switched. Suddenly, I was the good girl. As if she was living the life I was supposed to live, and I was living hers.

Charles didn't get flak for having a kid. Alayna did. Our mom, our relatives, our high school, and Alayna's college stacked it up against her because she had Isaac. How dare they. How dare they.


I took a long, deep breath. Then, I picked Alayna's ID out of my bag, swiped at security, and followed the signs to the advising office.

It was around quarter to four when I arrived. I was polite to the receptionist. "Hi, I'd like to speak to the advisor I spoke to last Monday, but I don't remember her name. Could you look her up?"

"Your name and ID number?" The receptionist asked with a taciturn why-am-I-not-off-yet inflection.

"Alayna Shaheen Sadiqqui, 2125736," I replied, holding my breath to hope that she didn't ask to see the picture on the ID card. The receptionist informed me "Last Monday you were with Mrs. Nunley in Office Four."

"Thank you," I replied. I walked to Office Four, Nunley. A porcine woman with a little green scarf wrapped around her neck and fashionably tied to one side sat with a huge poster behind her that read 'Expect Respect.' Brilliant advice. Two little clay bull figurines stared at me from her desk. I stood there for a moment to take a deep breath, reached to feel the knife was still in my pocket, and made eye contact.

"How can I help you?" She asked me as I walked in.

"Did you talk to Alayna Sheheen-Saddiqui about a Spanish class last Monday?"

"I'm sorry, who are you?" The woman responded, as I felt for the knife in my pocket.

"I'm her sister, and I'm here to tell you that she needs your help."

"Well, I can't discuss one of my students with you. She's entitled to privacy with her personal details."

"Actually, I already know the details," I replied coldly. "She passed every test with A's—"

"I'm going to stop you right there!" The porcine advisor cut me off. "I cannot talk to you about your sister."

"You failed to give my sister help, help she needed, help she asked for."

Nunley paused to let me speak, so I continued. "She demonstrated she knew the material by passing all five of her Spanish Exams with A's. She demonstrated full competency and proficiency in the material, and she deserves to pass the class."

"No, actually, she doesn't. Anyway, like I just said, I can't talk to you about your sister. Please shut the door behind you."

At that point, the overwhelming desire to kill her began to bubble hotter, and I got closer reaching into my pocket for the knife. "I'm sorry, you believe that you need to remain professional now, even though you told my sister – a teenage mom – that she shouldn't have had her kid."

"Back up or I'm calling security."

I ignored her, "You think that *that's* professional conduct? Aborting her baby is your solution to helping her pass Spanish? If you don't square this with me."

—"GET OUT OF MY OFFICE!"

"MY SITTO USED TO SAY THAT NO ONE ESCAPES JUDGMENT DAY, AND THAT MEANS YOU, TOO!" I was about to draw the knife, but it was not me who made first contact. The advisor grabbed one of the clay bulls and smashed it on her desk. It shattered into a million pieces. She stood shocked. I stood shocked. Before I could do anything else, two men rushed in and assumed I had smashed the bull, so they pulled on my elbows to immobilize me. They had no idea I had a knife in my pocket.

In between those men, I shouted. "She touched me first! She attempted to assault me! And she told my sister that she should have aborted her kid! I came here because she advised one of the kids at the college that my sister should have killed her son to pass Spanish! I swear! I swear she did!"

The crowd that had gathered looked at Nunley. Mrs. Nunley shook her head, "She's lying. Just get her out of here."

Mrs. Nunley looked at me, "You come back into this office ever again, and I will press charges," she threatened with the last word, and security guards dragged me off to a small room.

Nunley wanted the conflict gone. Apparently she was scared that I knew what she had said to my sister, so no one called the police, and no one knew about the knife. All that happened is that I sat in a room with the security guards who made me call my mom.

"Hello?" My mom answered.

The guard took the phone from me and explained the situation to my mom. "Yeah, it seems your daughter stole her sister's ID and broke something on one of our advisors' desks. She's not going to be allowed back on the premises." My mom assured them that I am a good kid and I never do this, and that she would fix me when we got home. The guard nodded his head like he believed her and said, "It's alright, kids mess up. Just come pick her up." I didn't say anything, but still thought how Mrs. Nunley deserved to die.

An hour in, my mom came into the office and performatively slapped me in front of the guards, which was meant to embarrass me, not to hurt. I looked down. We got in her car and went home.

We arrived back home, my mom toting four bags: office paperwork, home paperwork, purse, protein shake, and whatever else she carried. She set all her bags on the kitchen counter. I started to say "hello" to Isaac but she glanced icily at me, so I shut up. With utter stoicism, she grabbed Isaac from me and sat him in his chair in front of the television. She needed him to be entranced and silent. SpongeBob does that.

Our living room was the same space as our dining room, so I don't know why she made all the extra effort. Finally, once she finished sorting her stuff, she pulled out the seat across the table from me and demanded, "What happened?"

When I concluded my story, she confirmed, "So you didn't touch her? You just yelled close to her and she broke the figurine?"

"Yes." I had omitted the part about the knife.

"You're still probably going to get in trouble," my mom said without making eye contact.

"I'm sorry," I lied.

My mom waved her hand to shut me up. In blatant contradiction to the cultural norm, my mom sat stone straight. Other moms in the Arab diaspora would be sobbing about the dignity and reputation of the family. Without a father, our family was nothing more than Damascus ghetto, and my mom knew that.

She took her time and picked her words like produce. Finally, she began, "You know what my boss told me twelve years ago?"

"No."

"He told me that if I didn't respond to complaint emails within three days, I'd get fired. Even if I thought their complaints were stupid. Even if I was busy. You know what I've been doing for the past twelve years?"

I didn't say anything, so she prompted, "Do you?"

"What have you been doing?" I asked obediently.

"I've responded to their emails within three days like my boss said. Because that's my job."

I nodded.

"Did you hear me?"

"*Yamma.*"

"Do you know why I do that?"

"*Yamma.*"

"Why do you think I do that?"

I couldn't really come up with an answer. There was too much in my throat, and I was fixating on an icicle I could see outside the kitchen window.

My mom slapped the table in front of me hard enough to make my ears ring. "Are you listening to me!?"

Isaac began to cry.

I genuinely didn't have an answer for her. I stayed silent, staring at her. She slapped the table again. Isaac ran to the bedroom crying.

"WHY?" My mom yelled.

Every single interaction involved fighting and screaming and anger and hatred and threats of physical pain. As I saw her rear up to hit the table a third time, I stood up to get out of her arm's reach and shouted, "CAUSE YOU NEED THE FUCKING MONEY! WHY DOES ANYBODY DO ANYTHING? HOW THE FUCK AM I SUPPOSED TO KNOW THE ANSWER?"

My mom rose to face me at eye level. She stood taller than me because she wore platform shoes to work. "*Ihterami ummik. Bitehhkee ila ummik.*"

"Do you want me to answer you or not?"

"Do you have no respect?"

"I'm sorry." I said, lowering my tone to a whisper. "But how exactly am I supposed to talk to you?"

"*Jawabinee*."

"I don't know the answer!"

"Yes, you do," she spat. "But I'll tell you anyway. I respond to the emails," she said, calming down. "Because I love my family more than I hate my job. And that's the choice I'm given."

"*Yamma*," I listened.

My mom continued, "What you did for Alayna was not an act of love. You couldn't change the outcome. You just made it harder for her to get help from those advisors next time. In any case, she has to deal with what happens in her life. One class won't stop her. She is who she is, and God has given her a life of thousands of indignities, as he did me, and we both have to push through and learn the lessons life teaches us. You know that every single one of my mother's siblings stopped talking to us when I got pregnant with you, except Elias? You think I went up to their house screaming at them to tell them what they did was wrong? Of course not! It wouldn't have changed anything." My mom swallowed bitterness like tonic water, then reflected slowly. "The truth is you're too old for me to save you from your choices. Actually, I never could save you from them anyway."

So I didn't get in trouble for the knife in my pocket, but I never forgot the impulse to kill or the conversation which happened afterward. Isaac ended up raised by our mom. My sister became a nurse, marrying a man named Kareem, who was black, not Arab, having two more children with him, divorcing him, and ending up with another man named Javier.

After high school, I managed to get waitlisted then admitted to a good college where I met a short, thick man who studied statistics. He got a job in designing artificially intelligent chatbots and we married when we were twenty-four. It was my husband and the social status of my college, not the education itself, that comprised the crux of my upward social mobility. Married and in a mediocre white-collar job, I couldn't stop memories of that conversation from accosting me. In one moment, my boss needed me to search engine optimize a blog, and in the next my mom slapped the table and screamed for me to explain why she sent emails within three days. I called my sister and told her about that conversation.

"Really?" Alayna said. "I wasn't there. I didn't know you were carrying a knife."

"But what do you think about that message?"

"I mean, she's right, right? That's basic adulthood. You do the best you can with what you got," my sister said. And I realized that is the fundamental fact of maturity.

# BEAUTY IS NOT FOR MORTALS

Virginia Wolfe. Ernest Hemmingway. Sylvia Plath. Dead. Dead from their own doing. There's just something about writing that drives people to self-destruct.

Back in 2019 at AWP, the largest biannual writing conference in North America, I think I met fifteen clinical narcissists in one hour, as well as two drunks. Perhaps because writing is, essentially, an egotistical act. We only write if we believe that what we are saying matters. Whatever fundamental truth we're putting to the page must be filtered up through our minds and out our method of expression: a brush, a manuscript, or a dead chicken's wrung neck, like Ana Mendieta. We presume that we can make art.

Sometimes people try to claim that technical, expository writing is selfless—especially academics, who tend to be control-seeking compulsives. These types of writers obsess over provable material and technical description. Academics claim that what they are saying is objective truth, but their control and comprehension over their subject makes them the God of it, and that is egotistical. And, absolutely, that implicates me in writing this essay, in which I presume to be The Grand Judge of other writers. But what I'm saying actually is true, because I have yet to meet an expert, claiming objective truth, who didn't slide into pride of their own knowledge. Same with journalists, each of whom is the God of their stories—and when they realize that they are the God of their story, they claim they can also be The Judge, and they turn their story into Judgment as a Sellable Commodity. No writing—not even expository writing—is selfless. When you create something, you are in what you have created:

like your baby and your art.

Creative writers know the futility of the personal nature of writing, so instead of taking measures to mitigate it, we fall into it . . . and fall and fall. We collapse in on ourselves falling, only to resuscitate when we engage with the world around us. But when we return to our craft, we let ourselves implode further down, down, down, down. Imagine standing at the bottom of the ocean looking up at all the fish and then the sky. That's what poets see when they're looking up from the center of their soul, then they let themselves sink further down. As we fall, we begin to hear a voice that grows deeper, feel a tickle that turns into a burn, and we verge upon a comprehension of an ontological entity that we know is real but cannot fully sense: beauty.

I do not know why, but the search for beauty makes people want to die. It's as if God didn't mean for humans to know beauty—that beauty is too close to Him—so He sends guards to make us turn back. We feel beauty beyond our senses. We know that it's there just beyond our vision, and the pursuit of this ghost drives us mad. People who fall toward beauty live underneath reality, and it kills us. We implode.

While writers are living in our own crazy heads, other people are living their lives. Some people do not naturally lean towards artistic experiences, and they don't understand the value of reading fiction. But often these people manage to keep their house clean. Why? What does keeping your house clean do? It makes you feel better. Living in a clean house induces the feeling of serenity. When people clean their house, they are merely arranging objects in a specific way to produce an internal sensation. That is the beginning of sculpture art: the idea that we can put things into a specific order to cause a specific internal experience. If we clean our house, our minds feel clear. If we change the form and order of our things, we can change our state of mind, uniting physical reality with internal reality.

The museum in Germany which mourns the holocaust, Am Holocaust Mahnmahl, put up a sea of heavy concrete blocks in front of their entrance. At the beginning of the path, the blocks stand at eye height. But the path descends further down. As you walk through this exhibit, you verge on a sensation of panic and claustrophobia, but the path is small and there are people behind you, so you keep walking, and you feel like you're descending into a complete lack of control. This primes visitors for empathizing with the victims of the Holocaust. The objects create sensation very well. The concrete blocks in front of the Mahnmahl is probably the best modern sculpture art exhibit I've ever experienced. When you go to a museum and you see a painting, if you feel uncomfortable by the form, or you are induced to stare, or you see an emotion you've felt on a canvas, then you are engaging with art.

So, why does art matter? Why read fiction? Because it will explain something about your internal life to you. Art can help you find a critical change in perspective, meaning, sensation, outlook, emotion, or experience. In a well-written story, when the main character changes, the reader, who was emotionally attached to the character, will experience a critical change, a catharsis, a mimesis; a beauty. Art is the ancient practice of showing people their own internal lives.

In order to produce art, then, you must live deeply in your own internal life. You have to be real about what's going on inside of you if you want your art to help other people uncover what's going on inside of them. When you produce art, you have to stay inside yourself for so long that you begin to uncover the essence of your own creative power, and you become like God.

As the writer searches for beauty inside herself, she realizes that she can never get there. Beauty exits. We know that sometimes things are beautiful, like a sunset or the forest, but there is no single, tangible thing that is beauty. We can't source beauty. We can't trace it. We know it shows up sometimes, but we don't know its origins. I believe that the source of beauty is God, and that God did not mean for mortals to ever touch true beauty. I believe that God sends angels to turn us back in our pursuit of His beauty, and that those angels, in an attempt to preserve what is sacred to Him, make writers implode and want to die.

Beauty is not meant for mortals.

# ON JUDGMENT AS A SELLABLE COMMODITY

"Destiny, stop banging on that window!"

"Ms. Adelle, jump off a bridge and die!"

"Destiny, how do we express frustration?"

"Ms. Adelle, I'm frustrated you won't jump off a bridge and die."

Both you and my ex, Destiny.

Piercing through the burble of an industrial dishwasher, I saw Meredith stick her finger in her roommate's soup, twirling the broth with her fake red fingernails, instigating, maybe intimidating. "Mer! What do you think you're doing?"

"But she's not going to eat it anyway!" Meredith spat back, engaging.

"Yeah, I'm not touching it." LeAnn shouted back at me, goading Meredith to continue.

To be fair to them, the soup didn't look appetizing. It was made from a powder. But I couldn't tell them that. I ignored LeAnn and repeated, "Meredith, you are instigating."

"I am not instigating!" Meredith spoke over me.

"Yes, you are," I responded.

"You see LeAnn getting huffy and starting a fight? No. How are you gonna accuse me of starting a fight when there isn't a fight going on?"

These types of conversations could last hours if I engaged. "Meredith, get your finger out of LeAnn's dinner, and then get up and give her a new bowl, or I will write you up and you will not be out on Enrichment."

"Ms. Adelle, I'm just playing! Can't you tell it's a joke?" Meredith pulled her finger out of the soup. She pointed at the ceiling and licked the broth off her finger sensually like a lollipop.

"Is that acceptable?" I was only seven years older than this girl. I was pretending to act like her mom, channeling my sternest voice and coming off more as a know-it-all than an authority figure.

Meredith ignored me.

"Hey! I'm talking to you. Is that acceptable?"

Suddenly, LeAnn clicked her tongue at me, offended, and threw the bowl across the room into the trash can, yellow broth splattering up the wall.

"You better get that bowl out of the trash." I threatened. "Right now!"

LeAnn's eyes popped open as she realized how angry I was. She switched her empathy back on, and penitently retrieved the splattered bowl of soup, wiping it off the wall with napkins.

"Alright," I acknowledged.

As a youth specialist, I couldn't access the full history of the girls at Spruce Village, but I did know their reason for admittance to our program. For example, I knew that Lizzie was placed here after she, at the age of fourteen, performed fellatio on another child in her foster care home. Her judge deemed her a hazard to the other kids in the home and relegated her to residential treatment. She didn't need juvie, she needed help. From that, I can reasonably infer the abuse she had suffered prior to foster care placement.

Shannon punched her foster mom in the face. LeAnn squashed her auntie's cat with a book cabinet because the cat had scratched her. Lacie, Chloe, and Ana got sent here from juvie because they didn't have anywhere else to go when their sentences were done. Liberty chased her grandma around with a knife. Destiny masturbated too often in common spaces. Talia wet the bed, and jumped around like a minnow darting, and lied.

It took about two weeks to understand that my job would involve very little authentic connection or rehabilitation. If there were unlimited resources and the only goal had been patient rehabilitation, they would have gotten those girls as far away from each other as possible—this was the original impetus to end orphanages: the fact that traumatized people living in the same space exacerbate each other's problems. For example, when Talia told a particularly heinous story of abuse, it spread like gossip throughout the facility, and the other girls magically recalled repressed memories of men sticking beer bottles up them, too. They were not fully lying. They were collectively processing their traumas and creating a culture. Then a girl would be set off, start screaming, start fists, and the disruption of loud noises and threats would set another girl off, who flashbacked to time she was hurt with screaming and fists, and in the end, the whole facility would buzz like charged particles bumping around and no one knew who started it or what really happened.

If there had been unlimited resources, the ideal would be to put each one of these children in a separate home with a parent who can devote all their attention to the needs of that one child. But there are not enough people who are equipped to deal with a lying, violent teenager in their house, and violent teens reenacting their sexual abuse cannot be placed with young children, so residentials must exist as the next best option.

So, no, I didn't expect to be respected when I took the job. I knew it would be a hostile and potentially dangerous workplace. I took the job *because* I knew it was hostile and dangerous. Kind of in the same way a fatherless young man joins the military to grow up, or a woman, if she's conservative, embarks on a big mission trip or, if she's liberal, joins the Peace Corps. Those girls' behavior had nothing to do with me. I knew that. I was ready for it. However arrogant it seems; I believed that I might be of the stuff to face it.

"Ms. Adelle, I can put both ankles behind my head! Watch!"

"Ms. Adelle, can you write my name in Arabic?"

"Ms. Adelle, I'm reading a philosophy book! It asks if we ever step in the same river twice? What does that mean?"

"Ms. Adelle, let's play The Little Green Village!"

"Ms. Adelle, look! I passed Spanish!"


Lizzie listed, "Scissors, broken tile, plastic sheeting from a binder."

LeAnn chimed in, "I used the back of my earrings and a glass bottle I got out of the trash. You know, we were out."

I realized what they were talking about. "Stop it right now."

"What?" They both feigned innocence.

"Stop listing things you've both cut yourselves with. I know that's what you're doing. It's bad and it needs to stop."

"I'm making friends with LeAnn, don't you want me to have friends, Ms. Adelle?"

"Make friends some other way."

"We're actually sharing our traumas." Lizzie protested in a singsong voice intended to make fun of me.

"No, you're encouraging each other. A glass bottle you got out of the trash? Gross. This conversation ends right now!" I threatened with the sternest face I could muster.

They listened to me, but they pouted to make sure I knew they didn't like it.


"Come with me, please."

My boss, built like a stallion, and I stepped into the foyer with the metal detector wands. My boss cut to the chase, "LeAnn and Lizzy told Admin that you shamed them for cutting themselves. They said that you said 'gross,' and 'that's bad,' about their cuts."

I shrunk to the size of a six-year-old whenever my boss talked to me. Mrs. Regina commanded such a self-certainty that most people felt belittled when she spoke to them. If we worked at an office, Mrs. Regina would have been an insufferable boss, micromanaging and reprimanding us like we were the girls, but her supercilious comportment drove her success here at Spruce. When she shot a glance at the girls, they straightened up. They worried when she said, "disrespectful!" Mrs. Regina found the right place to use the gifts God gave her. When you work in a residential program, it's objectively very helpful to act like a dictatorial bitch. That is the personality type that succeeds in the role.

"You know they're manipulating the story." I clarified, "It would have been irresponsible for me not to have stopped it. I didn't call *them* bad. Lizzy said that when she went off campus for enrichment, she found a glass bottle next to a trashcan and broke it to cut herself. That *is* gross. She could have gotten hepatitis."

"You know those girls were going to go right to their social workers. You should have seen that from a mile away."

"Should I write an incident report?"

"Under no circumstances should you document this conversation. If you put their claims in writing, that's a great way to force me to pursue disciplinary action against you." Her use of the word 'disciplinary' didn't help with my feeling of being talked-down-to. My boss saw that I was only a couple years older than them, and she probably thought of me as more like them than like her. But Mrs. Regina protected her staff. Mrs. Regina explained, "You know how Admin will be."

I exhaled. "All right, what would you have said?"

Mrs. Regina sighed and explained, "The response is, 'we're not talking about that right now,' and if they keep talking about it, you threaten to give them a pink slip for not following directions. That way you're not in violation of any of the new policies Admin keeps changing."

"All right."

"All right, good." My boss agreed, "No further discussion. I'll submit a paper saying I do not know what happened in your conversation, but that we train our employees on sensitive language."

We left it at that.

The topic of inclusive language juiced up hatred between the floor staff and the legions of supervisory social workers who sent us dictums down the pipeline. The girls loved their social workers. Their social workers took them out and bought them what they wanted. But if a girl behaved badly in a social worker's office, floor staff were called to corral her back onto the floor, always screaming and sometimes kicking. Yet, the legions of social workers decided what we could and couldn't say?

The truth is, yes, self-harm is wrong. It's wrong because violence against ourselves is still violence. The psychological complex behind violence doesn't excuse it. As it is not okay to violate another person's body, it is not okay to violate your own body, because your body is the body of a person. An uncomfortable but important truth behind the psychological complex which drives people to self-harm is egoism. If a girl is okay with hurting herself, but she would not hurt another person, she's implying that she's made of a different stuff than the people around her. Basic human rights don't apply to her because she is trying to convince herself that she is not merely a basic human. There is remarkable similarity in the pathology of a girl who thinks that she's better than everyone else and a girl who thinks that she's worse than everyone else. If they are too caught up in their own *I Am*, then they will never successfully connect with the people around them, and they will rot in their own loneliness. But apparently trying to explain this to my girls is insensitive to their mental health.

While I was still on edge after the conversation concerning Lizzy and LeAnn's cuts, Meredith had a bloody line over her stomach on her shirt. We took her into the private office and asked her to let us clean the wounds, which meant she had to take off her shirt.

"No fucking way, you perverts!" Meredith screamed.

Mrs. Regina sat behind the desk in front of file cabinets with enrichment supplies like paints and Crayola crayons, which the girls hated anyway because they were too old for arts and crafts. I stood in front of the doorway to make sure Meredith didn't bolt. Mrs. Regina calmly explained, "Meredith, I need to see why your shirt has a blood stain on it."

"You just want to see me naked!"

This was terrible, and I was glad a more experienced person was there to navigate this situation. Mrs. Regina soothed, "No one's angry at you. You're not in trouble."

"I'm not fucking stripping for you! There's a camera in this room!"

"Meredith, you need to put a bandage on." It wasn't a cat scratch wound. It was a big, festering, oozing cut.

"Fucking pedophiles—that's what you fuckers are! I bet you're going to watch the footage over and over again and jack off to it! Motherfucking perverts! I hate you! I hate this place!" Slinging abuse was their first line of defense. Every day I went into work, the girls said stuff like that to staff. Every day.

In two years, Meredith would be eighteen, and there would be no staff to fight with her about cleaning her wounds. If she didn't clean them, they could get infected, and she would endure the wrath of reality by herself. A good part of me wanted to let her endure it right now. Then I looked at her and I saw a child, and I couldn't be angry at a child for acting childish.

Meredith was correct that cameras covered every square inch of the building including this room, except the toilet stalls and showers. At least Mrs. Regina's boss and a standards representative would have to watch the video of her stripping to ensure that we conducted ourselves profession-ally during the medical intervention. Spruce said on its website that it was an equal opportunity employer which didn't discriminate based on sex, but that was just a principle. A home like this just shouldn't have men in it, and we all knew it, including the men. Even if potential male employees earnestly wanted to help, their presence disrupted the home. And if they handled situations like this, they made themselves look very suspicious.

Mrs. Regina ignored Meredith's antagonizing comments. "We need to help clean your wounds. We'll let you clean them if you do it properly."

Meredith continued shouting strings of slurs and abuses, spitting and crying. The idea of having to take her shirt off clearly humiliated her, but saying no one would judge her didn't help her because she al-ready judged herself, and there was no trust enough for me to alleviate that. She didn't trust me.

"I'm done arguing with you." Mrs. Regina started. It was an impossible decision. We could either forcibly strip the clothes off her, violating her body, or we could let her self-harm wounds fester, violating her body. Mrs. Regina stepped outside. "I've gotta call Admin." She beckoned Ms. Elena to back me up as she stepped out.

Meredith started crying and banging the walls. Every time I anticipated a physical altercation, I worried that I might lose my temper this time. If I swung first, legal action would be taken against me. If I swung at all, I probably would be fired for liability reasons, but legally they couldn't convict me of assault. I reminded myself that I could probably win against morbidly obese Meredith, who was literally drowning in her own fat. I could just run in circles and win against her. Between the side effects of antipsychotics and the compulsion to turn an abused body into a sexless creature, most of the girls at Spruce pushed two hundred pounds.

The facility fed them as healthy as they could. Healthy like Americans eat. A serving of green beans next to their corn dogs, Honey Bunches of Oats instead of Fruit Loops, and peanut butter instead of Nutella, even though American peanut butter has about as much added sugar as a European candy bar. We fed them appropriate portion sizes. Didn't matter. Most of the girls waddled around, knees knocked inward because their bones buckled under their weight.

Meredith banged the window again.

"STOP IT!" I screamed at her, and she spun around scared of the force of my voice. "If you go to the hospital, guy doctors will treat you." I explained. "Here, you can wash it yourself. Just, come on."

"Fuck you, Ms. Adelle." Meredith replied as she sat down in a chair. But at least I had gotten her to calm down. Then, by a miracle, she said, "Alright." She took off her shirt and I saw two huge breasts bleeding just under the nipples from wounds that were clearly made with a knife. I held up a mirror for her. She applied Neosporin herself. The entire procedure was invasive, embarrassing, and terrible. When we were done, Meredith stormed out of the office and shouted to all the girls in a communal space we called a living room. "Ms. Adelle felt me up!"

"I am strong enough to do this," I reminded myself.

It is statistically true that impoverished people commit violent crimes at higher rates than other socioeconomic statuses. Liberals in California and Washington have used that correlation to justify expansive social reform, believing that if they eradicate poverty then they will eradicate violence. This thinking is terribly mistaken, and it becomes dangerous when it is used for the rehabilitation of individual violent people.

Even if that statistical correlation is true—and I have reservations about that, because I think that it is more likely that impoverished people are convicted at higher rates than it is that they commit violent crime at such a disproportion, but even if they did—talking about systemic injustice and the cycle of poverty makes those girls at Spruce feel hopeless. My girls cannot solve poverty. If someone tells my girls that their problems are caused by poverty, which they know they cannot solve, then they walk away thinking they cannot solve their problems. That's what happened every time a stupid college student or, worse, a social worker came in to facilitate dialogue. Their explanations strip the girls of their agency. It imparts hopelessness. The academics and the professorial class who keep talking about 'poverty and violence' looking down from ten thousand feet in the air have no clue about the impact of their words on the people suffering from poverty and violence. Newsflash: it keeps them in their own heads and hurts them.

Alternatively, if you tell a girl that her problems were caused by evil, she can solve it. She can conquer the evil in her own heart. If she believes that she can end the cycle of abuse she suffered by becoming a better person, then she can exert agency over the shit in her life. She can get better. Rehab, AA, and correctional facilities emphasize personal responsibility to help their patients use agency.

When we got upstairs, my coworker Ms. Elena and I stripped Meredith's room of everything except a pillow and a blanket. All her artwork, little rocks she liked to collect, fancy soaps, and her makeup—which she loved most in the world—got rounded up. As we took her makeup, she screamed and grabbed the bag out of my hands. "How am I supposed to cut myself with mascara?"

"You can't have anything if you're on watch," Ms. Elena explained.

"If I want to cut myself, I will cut myself with anything! So, there's no point in taking my stuff."

"You can't have anything if you're on watch," I repeated. I attempted to take the makeup bag from her, but Meredith elbowed me in the face as she screamed. "YOU'RE NOT TAKING MY SHIT!"

I clutched my left jaw and backed out of the room, making sure I had eyes on the escalated client. "I'm strong enough to do this," I reminded myself. Meredith banged on the walls, crying. She overturned her bed frame. She ran out of the hallway into the bathroom and ripped the shower curtains off the bar, spraying little plastic shower rod rings all over the bathroom. Procedure dictated that if the client was breaking property, we let them. Things can be replaced. People cannot. So, if the client can vent their emotions on shower curtains, then by all means.

Meredith exhausted herself and went back to her room to sob. "Nobody cares about me! Nobody cares!" However, even though she just elbowed me in the face, I wanted to let her know that people do care about her. I absolutely wanted to love her. I wanted her to be safe.

I shouted down the hall, "Yeah, Meredith. We're here helping you, feeding you, forgiving you, keeping you safe because we don't care, mm-hmm. You're so right." These girls needed a nonviolent, steadfast authority figure with their best interests at heart—someone who had already killed their ego enough to endure the relentless string of abuses the girls threw without taking it personally. Someone with complete temper control. They did not respect someone who they could control. They feared those they did not control. They tried to control you with everything they had. They had no control over their lives, nor should they.

Meredith wanted more compliments, so she got up and slammed the door, which was a boundary violation. The door should always be open, and, given this policy, it was annoying that the facility had doors on their rooms at all. Ms. Elena and I approached Meredith's room and informed her that we were going to open it.

"No! Please! All I want is some privacy!"

"I understand you're upset, but this door must stay open," I repeat.

"You need to shut up before I can't control myself and kill you. Nothing matters, anyway."

"That's not true." I shout through the door. It's very weird to be so inured to abuse at my job that a violent teenager can threaten to kill me and I still have an urge to de-escalate, soothe, and comfort. "Many things matter. Like, for example, your behavior matters. Others will respect you, and you will learn to respect yourself, if you behave like a respectable person. If you want to respect yourself, you have to be respectable."

The door slowly opened. "What do you mean, Ms. Adelle?"

I was surprised at the calm in her voice. "I mean that if you want to respect yourself, you have to act like a respectable person," I stammered.

"Like what?"

"Like, if someone said I had a big nose, I might feel bad for a moment, but then I would look in the mirror and see a button nose and get over it. But if someone called me fat, it would really hurt me because my thighs are still very thick and I've struggled with my weight my whole life. Do you understand what I'm saying?"

Meredith opened the door completely and calmly. She furrowed her eyebrows in connection. "Like, the things that hurt us the most are true?"

"Yes!" I encouraged. I had finally connected with her and conveyed something that might actually help this child, so I continued. "And the first step is admitting that. If you say sorry and really mean it, it will recalibrate your soul, Meredith. Just try."

"I mean, yes." Meredith admitted. Meredith looked down and then back up at me. "I'm sorry that I elbowed you, Ms. Adelle." For a moment, it was beautiful.

"I understand that you felt an immense lack of control over your life when you did that." I responded, but that pushed it too far. Meredith switched back to being angry. She shut the door on me. Poof, gone.

Meredith flew down the stairs and out the fire exit. She set off the fire alarm, then parked herself at the picnic table in the backyard of the facility. My jaw still smarted from her elbow, and what was supposed to be a beautiful moment didn't actually materialize. No staff was about to go out there and get into another fight with her.

Twenty-two out of the twenty-four girls at Spruce Village took meds. Sixteen out of twenty-four girls took antipsychotics, even though their problems were based in trauma, not psychosis. As trauma gets more extreme, the line between trauma and psychosis blurs, and the medication showed it. We're talking about a hundred milligrams of Lexapro per night plus a neuroleptic. Latuda, Clonidine, Zoloft, Trazadone, on and on. Two of our girls were completely dependent on medications for sleep, and we kept giving the pills to them anyway because if we discontinued the drug, then the girl would be up until 4 A.M. every night for five weeks after, and our facility couldn't manage a child going through that severe of a withdrawal.

Before I worked at Spruce, I didn't know how to spot the effects of fetal alcohol syndrome. But then I noticed that the girls who were medicated with all four categories of drugs we gave: amphetamines, antidepressants, neuroleptics, and anti-anxiety, tended to have rounded, bulging eyes and thin upper lips, like Meredith. That's how to spot fetal alcohol syndrome. The girls didn't know this, but we designed their shower groups based on their medication profiles. That way, if the girls tried to swap their meds in line, which they especially did with amphetamines, they would be more likely to swap with other girls who were prescribed the same drugs. I anticipated that eventually some girls would be cut off from their medications and turn towards illegal drugs, but based on the amount of drugs I gave those girls every night, that thought didn't seem to have occurred to their psychiatrists.

Once all the groups had washed and taken their meds, I went down to check on Meredith, who was still sitting on the picnic table in the yard. "You ready to come in now?" I called.

Meredith turned around to look at me, but her eyes were glazed over. Blood streamed down the arm I could see. I ran to her. In her right hand, she held one broken plastic shower rod ring. On her body, there were more than two hundred small gashes, and collectively, the blood streamed down—and there was one long, vertical, deep cut on her left forearm tracing the vein. I called Ms. Elena from my flip phone.

"Hello?

"Bring gauze, Meredith attempted suicide."

"I'm the only staff upstairs. I can't leave."

"Fuck, all right."

I used Meredith's sweatshirt to tie around the deep gashes in her forearm. I called an ambulance. Mrs. Regina hurried out, toddling like the elderly run, with gauze. Mrs. Regina bent down and kept wrapping up Meredith—wrapping and wrapping until the girl looked like a mummy—as if she was already dead. Meredith had cut her whole body except her face. The front of her legs, shins, and thighs, stomach, chest, fingers, toes—every inch of her was covered in small cat-scratch wounds. She must have planned to bleed to death by small wounds before mustering the courage to give herself the big one. An ambulance arrived. Police came.

A fat policeman reprimanded us, "You let a girl sit out here for an hour doing this? What the hell is this place?"

On official documentation, Spruce stood accused of neglect because we allowed the attempted suicide to happen. Meredith was stabilized at the hospital, but that wasn't a good enough excuse. The police contacted Child Protective Services to investigate. On local television, a woman in a red dress reported, "Breaking news! Spruce Village Family and Foster Care Service's group home is facing heat when it was found that staff failed to respond to someone in their wards attempting suicide for almost an hour. Chester has the latest." The news channel cut to the anchorman named Chester, standing in front of the entrance to our campus reporting that a minor whose name could not be released attempted self-harm and that it appeared that the minor had the opportunity to cut themself over two hundred times before staff noticed and stopped them. He said 'it appears that' to absolve himself from telling a lie.

I created this terrible, embarrassing, public thing? I caused it? I opened Pandora's box and released a writhing black horror story into the air? This wasn't my fault. I was following procedure! The procedure I had been taught was that we do not pursue hostile clients, and we do not call the police unless they are off grounds. I did what the facility told me to do. When I realized she was hurt, I tended to her wounds, like I had before. And what the hell? I chose to do this job because I wanted to do good. I wanted to be good. I wanted to help these girls.

If Spruce Village needed someone to blame, it would be me or Mrs. Regina. Mrs. Regina couldn't see what Meredith was doing through the grainy camera footage. Her or me. I'm low on the chain of command. Only my immediate family would care if I got fired. It didn't matter if they lost me because they would hire a new woman who also wasn't good at college to replace me. Mrs. Regina had twenty years of experience, but that made it harder for her to claim that she didn't know what to do. If Mrs. Regina got fired, this place would descend even further into chaos and trauma because a worse supervisor would use worse judgment. I began feeling protective of Mrs. Regina and considered taking the fall for my friend. But then again, if my CEO took enough heat and realized she needed to fire one of us, you bet I would point out it was Mrs. Regina's job to watch the security footage. But it wasn't either of our faults! I was instructed to not pursue hostile clients in training. Mrs. Regina couldn't see through the grainy footage what was going on. We tried our best with the resources we had. We followed procedure. Everyone who had ever worked as a Youth Specialist at Spruce knew that no one could have done it better than us. But everyone outside of the facility believed that they could have done better.

To be clear, the administrators investigating what happened, the psychology PhDs—such poor excuses for intellectuals—talked about trauma-informed procedures, and they believed they knew what to do in this situation. They dictated our procedures and told us how we should have handled Meredith. But you know what they never did? They never handled Meredith. They never wrapped a girl's wounds. They never got elbowed in the face on their shift at work. And yet those administrators were the ones empowered to set procedures about what to do with hostile clients; they blamed us, and they fired us.

The general spectators of the news also believed that they could have done it better than us. No question in their mind. The people eating Lean Cuisine in their living room who learned what happened through a reporter, who didn't see the care and forgiveness I tried to show her, all muttered at their televisions that Spruce Village staff need to get off their asses and pay attention to their jobs. The people never would have to test what came out of their mouths. They all believed that they could have done it better than us because they would never have to test it. They weren't there. They never tried.

What is that cognition? What is this obvious knowledge which lets you feel so sure that you would have done better? You never would have left an orphan to hurt themselves in a yard. Preposterous! You would have been able to save Meredith.

Well, there is a line of logic to it. It goes like this: when we read the story, we recognize its evil. We quickly remember that we are not evil. Therefore, we believe that we could have done it better. That is to say: we believe that we are good, and that the story is not good, so we have every reason to be moral over it. That cognition is called self-righteousness, the belief that you would have done better because you're inherently good. Self-righteousness is not righteous. Self-righteousness was what led me to take this job: the preposterous belief that I might be of the stuff to handle this pain and trauma. But that same belief is what smug people feel when they read about someone else's screw ups in a news report.

Self-righteousness is what biased news articles sell, like the kind the anchorman Chester was selling about what happened with Meredith. When you read a story with obvious evil in it, you get to be reminded that you are good—that you wouldn't have done that, that you could have done better because you *are* better. When you're done reading a sensational article, you feel more certain of your own goodness and therefore do not question the badness of the report. Judgment is a sellable commodity. When you read about other people's mistakes, it gives you a false sense of assurance that you would never make that mistake.

When the news reported that Spruce Village had let a girl bleed for an hour, the reporters weren't technically lying; they were just telling an incomplete truth to help the reader spot the evil easily. This would titillate the reader's ego and spur the reader to keep reading. This is the game of propaganda. The industry that sells us self-righteousness is propaganda. When you read sad, incomplete, biased stories in the news, they are selling you a feeling of self-righteousness.

And, no, there's no resolution for those girls. I couldn't connect with them, and there was no definitive moment of healing.

But I tried.

# WHY THE IDEA, "IT IS NOT AS IT APPEARS TO BE" IS THE FOUNDING MYTH OF WESTERN HISTORY

remember sitting in Humanities of Western Civilizations class as an eighteen year old, half-blinded by harsh auditorium lights. In my cramped little seat, I stopped listening to the words of the professor because I had noticed that the story of Western History is told in a certain order: idea, civilization, idea, civilization, idea, civilization. Or, if you prefer, the order: idea, culture, idea, culture, idea, culture.

That is to say: Chapter One of the story of Western History began with a man sitting cross-legged in a cave saying, "Imagine your senses are actually shadows on a wall, and then you will begin to see things as they are." This was Plato's idea of the Form, and most of the time the story of Western civilization starts with that idea. Other cultures have different foundational ideas, but that one was the West's. Shortly after that idea was written down, Rome was built. Chapter Two in the story of Western History was Rome. At the height of Rome, a new idea came—the idea that morality is found in humility, not in conquest. A Messiah came down and separated the idea of 'strong' from the idea of 'right'—this is what Nietzsche meant by 'a slave revolt in morality.' Chapter Three was the ideas of Christianity. Chapter Four was the Middle Ages. Then people began moving their thoughts around like numbers and calling it 'logic,' and so Chapter Five was the Renaissance and the Age of Enlightenment. After the Renaissance, Chapter 6 was mercantilism, colonialism, and the Industrial Revolution. Chapter 7 was spastic on-off, zoom-zoom, postmodern thought. And Chapter 8, starting in the year 2000, marks the dawn of the Digital Age. I'll say it again:

Chapter 1: Plato's idea of the Form

Chapter 2: Rome

Chapter 3: Christianity

Chapter 4: Middle Ages

Chapter 5: Renaissance and Enlightenment

Chapter 6: Mercantilism, Colonialism, Industrial Revolution

Chapter 7: Postmodernism

Chapter 8: The Digital Age, starting in the year 2000.

Do you see how that goes: idea, civilization, idea, civilization, idea, civilization?

For full disclosure, I am aware that the pattern I see probably has more to do with the pedagogy of my class than the reality of history. Pedagogy is a fancy word meaning *how you teach things*. As in, if you've ever had a professor who knew the material but couldn't explain it well, that teacher had bad pedagogy. My humanities professor had to cover a massive amount of history very quickly, and she designed the course so that it was likely that most of the students would get something out of the course. But I do not think that she intentionally designed the course to go: idea, civilization, idea civilization. At least, I do not think she was conscious of that pattern, because she never expressed that idea. I think it simply made sense to her to tell the story that way because there is some sort of relationship or interchange between ideas and civilization. So, what came first? The idea or the civilization?

Based on my education, I would say that most academics and published writers nowadays think that ideas come from material circumstances. They trace ideas backwards in history. "Metalwork with bronze brought Greek philosophy because metalwork preceded writing philosophy down." This belief is not just important for their conjectures on the patterns of history. This belief influences other aspects of their political and social mores. People who believe that civilization gives birth to ideas might call themselves 'Materialists' or 'Marxists' and they tend to say things like "People think this way because this happened to them." "People are trapped because of their experiences." "People are trapped in their economic circumstances." "Economic structures cause crime." "A person born in low circumstances has no agency to get out of it." "The proletariat can never lead rich and fulfilling lives because the definition of fulfillment is loving what you do." "A good social structure must be in place before we come up with good ideas." "Object then idea." "Object then idea."

I am not entirely convinced that "object then idea" is the correct order. At minimum, I think it is possible for ideas to start civilizations because ideas are more powerful. Also, Materialists and Marxists tend to be stuck in an infinite loop of bad logic because they do not realize that their belief, that civilizations spur ideas, is itself an idea, and that their political movements to design society are flowing from their heads to their politics to the world. You see how that flow goes? "Idea then civilization."

It is true that an experience can change an idea, like poverty causing someone to hoard objects in their house. It is also true that that can apply to macroscopic trends in which social structures change consciousness, such as factory work causing timeliness to be a value where it wasn't before. But the reverse is also true and much, much more powerful. Ideas can color, frame, and animate experiences. What's a bigger change: living in an electric house or understanding electricity? The fact that I live in an electric house does not mean I understand electricity. The fact that someone understands electricity is the reason I live in an electric house.

Stay with me: in the 20th century a new school of philosophy started to develop called 'postmodernism.' Postmodernism begins with Nietzsche, but it took full life with people like Derrida and Foucault. One tenant of postmodern philosophy is subjectivism, the opposite of objectivism. Subjectivism is the idea that things that are observed are done so in a man-made system, and you can really change the way you observe things by changing the system you're observing them in.

The field of science was originally founded on observing things, objectivism, but once people began thinking about really really big things, like outer space, and really really small things, like quarks, people had to depend on logic outside of observation. In quantum mechanics, electrons behave differently when you observe them.

Do you see how that scientific principle correlates with postmodern subjectivism? The idea that electrons change when you observe them parallels the idea that the system controls what you are observing. The way electrons behave is a wonderful metaphor for subjective philosophy. In a way, postmodern philosophy can be thought of as man's best attempt to incorporate the way electrons behave into their theory of how things are.

- The double slit experiment was performed in the early 19th century.

- Postmodern philosophy began in the early 20th.

- Computers were developed in the second half of the 20th.

- A society with the values of postmodernism, The Digital Age, began in the 21st.

Do you see how the idea of postmodernism preceded the Digital Age? The order went electrons, postmodernism, new culture. Smart Materialists who actually understand materialism, like my friend Georgiana or my fiancé's friends Hernan or Wei-hoo, would say that the most important thing in that line is the electrons themselves because they are the beginning. But I think that postmodern philosophy is the great leap forward in the story of humanity. And there is a third group, confused Materialists, who think that the order is electrons, new culture, and postmodernism.

It is true that the children of elite people in the 21st century are taught to read Sartre, Foucault, Nietzsche, Marx, Freud, and Derrida. Maybe Wittgenstein, but I never read him. When poor people imagine wealthy elite people, they might imagine a woman with pearls on a red couch going to the opera, but that is because they do not interact within systems of wealth and power in America and have no clue what they're imagining. People who are powerful in 21st century America don't love the West, and they wouldn't, broadly speaking, value or prioritize an art form as Western as opera. They are rich from many different countries, who own trade routes protected by the US Navy, and they care about multiculturalism, relativism, and subjectivism. If you want to act like an elite person, study data collection and global trade routes, learn to speak many different languages, only eat organic food, and read postmodernists. The leaders of The Digital Age pay hundreds of thousands of dollars for schools to teach their children postmodernism, and that is not a coincidence.

I believe that ideas come first. Something in someone's head turns into something that can be touched. Ideas can bring material circumstances into existence. Ideas can be traced to material, in the same way they say God willed the universe into existence. On a much smaller scale, human ideas have the same ability to materialize, to incarnate, to vivify. And I believe it's happening right now.

To provide another metaphor, I will briefly mention that the question "What came first? The idea or the civilization?" can be understood as a more abstract version of the debate "What's right? Science or religion?" In this analogy, the stance "civilization brings ideas" substitutes for science and the stance "Ideas bring civilization" substitutes for religion. Or, "What's a greater power in the universe? Creative power or the order of things as they are?"

If you're confused by that point, it's probably because you have not been exposed to the history of people fighting about science versus religion because you live in an age dominated by science. This debate is not relevant to public discourse anymore. But it raged for about three hundred years, between 1600 and 1900. Isaac Newton articulated the ever-present and observable tug that objects exert on each other, which he called gravity, and this way of thinking became contagious. People stopped connecting with the ever-present force through ethics and morality (Christianity) and they started connecting with the ever-present force by playing with objects, and this idea caused them to stop centering their interpretations in beliefs like "I am subject to the mercy of the ways of the world," to "The ways of the world go boom zip boom, so if I boom then it will zip." I am not claiming that you can historically trace people consciously deciding to do this. It is merely a fact that in the course of those 300 years, people replaced religion with science, and the ever present force which occupied the thoughts of the times went from God to things like gravity.

So, going back to the question if ideas will civilization into existence or if civilizations goad ideas into existence: the line of logic which science must follow is observation of material things, then extrapolation. Observation, extrapolation, principle. The line of logic in which creative ideas flow is idea, reality. Idea, reality. That's all. Idea, reality. Person, mimesis. Way, is. Creativity is a much more ancient way of understanding. So what is creativity? What is the religious side to this debate? Here's an example:

Imagine a yellow circle,

Moving in an arc

Deepening into a ball

burning hot

and thinking, "that is the sun."

That is how we first saw the sun. That is first consciousness. If a guy asked his friend "What's the sun?" the friend would say "It's a hot yellow circle that comes up in an arc." That's what the sun is. That's what I sense, and therefore that is my thought about it. Then the friend asked, "What is water?" And the guy said, "It is cold in my throat and salty in the ocean and it comes from the sky." Those are ideas. The guy is thinking about the sun and water using thought to come up with the mental image of a large, hot, yellow circle. That is first consciousness.

It took a lot of really smart people, whose brains were just the same developmentally as ours today, to realize that the sun could be thought of in a different way than as a large, hot, yellow circle. Someone had to realize that the sun was a star. Before someone had that idea, humanity had always separated the sun from the stars as a fundamentally different existence. Sun, moon, stars. Rock, water, air. How could the sun be a star? This is not an experience shift. This is an idea shift.

And that idea shift pertained to other aspects of life besides the solar system. A guy who thinks of the sun as a star will ask, "Why does it look so big?" And he will think about how much closer we are to the sun than any other star. He will visualize how that looks, and how that weighs, and consider that size and weight are related, and that weight is a force of movement, and ask himself what gravity is. That is the creative side to that story.

So what is creativity? What is the religious side to this debate? Why is the idea that *things are not as they appear to be* so important that it is the beginning of the story of the West? Because it links you from civilization back to ideas. What Plato expressed is the essential move you need to make if you're traveling from "what is" to "what could be." It is an articulation of the creative process.

The beginning of the story of Western Civilization has to start with "Imagine your senses are like shadows," because Plato needed to move people beyond first consciousness. The senses would never tell you that the sun is a star. Only creative revisioning of the senses could suggest the sun is a star. Only creative revisioning could come up with subjectivism, making our thoughts move like electrons, creating new order.

"Things are not as they appear to be," is the eternal fact of creation. It's the mental process creative people follow when they generate new ideas.

## POST SCRIPT:

Warning! There are some very deep thinkers who are also sociopaths, who understand that ideas have the power to will civilization into existence, but who use false ideas to do this. Those people often work in government and media.

## EXHIBIT ONE:

In 2015, with acne-splattered faces, Guldehan's Driving School paired Hannah and I together as road partners. Neither of us wanted our license, but we knew that we had to get a license eventually. The aforementioned Guldehan, a Muslim entrepreneur with eight daughters, put a second brake in the passenger seat of her car and managed a profitable business. Guldehan taught many kids who lived north of my high school how to drive. Hannah and I watched each other run into trash cans while Guldehan switched between patience and her wit's end.

Admittedly, I was slow to learn driving. Guldehan kept shouting, "No! Stay in your lane! What is this? No! You see the sign?" Although it is true I wasn't a good driver, the way Guldehan yelled at me made me feel demoralized and embarrassed behind the wheel.

Hannah noticed this. "She *is* staying in the lane," she said in a moment of solidarity. I remember that.

Hannah was observant, pensive, and interesting. Hannah and I both aspired to write. Hannah laughed at herself when we were studying for the exam together. She dressed in bright colors. She crossed her ankles when she sat. Even though we only fostered a situational friendship, Hannah came across as funny and attentive to the needs of the people around her.

The last time I saw this person, my friend deadnamed them.

"Hannah?"

"It's Tobias," they explained. Bulbous, infected acne, but worse than the kind they had when they were a teenager, pocked Tobias's face. It looked like testosterone acne, but I wouldn't ask if they were on T because I had watched enough TikTok reels to know that everyone knows that's an invasive, personal question unless you're also trans.

I visited their website. Apparently, Tobias cranks out novels like Stephen King now. I read Tobias's bio and self-branding—"trans, non-binary, autistic historical fiction writer." Their branding does not argue that their writing is persuasive, or the content is raw, or the philosophy is meaningful. Instead, they claim that their books are valuable because a trans, nonbinary, autistic person wrote them.

The autism claim especially confused me. I know that I'm not a doctor, but I would have noticed if Hannah was autistic. I spent many hours trapped in a car with this person. When I knew this person, they were empathetic and emotionally available to me. If this person wasn't autistic at the age of sixteen, why would they have suddenly become autistic at the age of twenty-three?

## EXHIBIT TWO:

I once needed to speak with a professor, so I walked into her office. Posted on the frame of her door was a picture of an infant in a crib. I asked, "Is this your grandchild?"

Whatever exchange we had, she was taciturn and kept looking down. Eventually she took me into a different office where the department administrative assistant worked. Once she had an audience, the professor turned to me and said, "How dare you ask me if that kid is mine! You didn't come into my office asking me about my work! You immediately asked me about my reproductive capabilities. You only like me because of my uterus! You see this office?" She pointed at the department chair's office. "It's almost always filled by a man. And most of the time, an idiot man. And I will never..."

I said, "I'm sorry. I'm very sorry, please." I tried to get her to stop yelling at me.

She apologetically admitted, "I think that I've scared you, but..." and then repeated all the reasons she's justly angry.

I said, "I'm sorry."

Later that day, she sent me an email about it. I ignored the email.

Two weeks later, she sent me a follow up to continue this conversation. I said my grandma had been diagnosed with cancer.

She said I should be checking my emails anyway.

Eventually I told administration.

It turns out that 1) she did have a child and 2) that child was not the child she had posted in her office. She had, instead, decided to print out a picture of a baby she found on the internet and post it on her office door because it symbolized something arbitrarily specific to her.

## EXHIBIT THREE:

On the first day of class, coffee in hand, a different professor from Exhibit Two told our class, "I will work you in this class harder than any other class in this college. You will be silent when I give you feedback on your writing. If my feedback makes you cry, I don't care. No one wants to see that. Wear a hat."

If I had common sense, I would have ended Exhibit Three at the first paragraph. Instead, I'm going to tell the whole truth, even if it gets me in hot water, because the truth is worth telling.

There are many unhinged professors who act like that, but generally if professors berate and humiliate students enough, Administration will rein them in. Professors that are published in prestigious journals have more leeway, and this professor had been published in many top tier journals. However, even if you're prestigious, most of the time there is a cutoff of common decency somewhere. I had many conversations with fellow students about this professor's communication style and general demeanor, and over and over again the conversation reached the glaringly obvious conclusion that, due to the political climate of the university, the administration would not cite or stop this professor because she was protected by her race.

## EXHIBIT FOUR:

One of my friends from college is a half-South Asian man who is Muslim and Queer. As a practicing Muslim, he had a job as the House Director of Jewish student housing on campus, impelled by his mission of bringing greater understanding between Muslims and Jews. He identified as nonbinary. He was accepted to an Ivy League's Religion School for his master's.

I knew this person before he earned his fanbase of Northeastern Socialites. I knew him from Arabic class when he was just learning to respond to Middle Eastern media. An extremely famous Egyptian talk show host named Shiekh Sha'rawi was quoted often in this class. This talk show host believes that Islam is the answer to all evil and all suffering, and that, among other things, a woman's dignity, religious transcendence, and personality are best expressed when she is in complete domesticity.

Working is no place for a woman, he believes. There's no compatibility between that belief and contemporary Western feminism, so it was hard to smush a clip saying a pregnant woman should be in the kitchen into the West's general theory of diversity and inclusion.

My half-South Asian friend, at first, tried to incorporate Sheikh Sha'rawy's view into his own to authentically engage with a widely circulated voice of contemporary Islam. Then, my friend realized that Shiekh Sha'rawi could never be respected by or assimilated into the belief systems of secular white people, so he had to cast him off. I watched my half-South Asian friend scrub his religion of anything which people who run universities find offensive. Essentially, I watched him learn contortion. And that move worked wonderfully, as evidenced by his elite education and burgeoning career.

## EXHIBIT FIVE:

"Motion to open discussion on the topic of piece #32?"

"Object!" Arden would say.

"What's wrong?"

"Piece #32 is ableist. It references Joseph Conrad's mentally disabled character as a retard."

I liked the piece. "That's because Joseph Conrad called that character a retard. The paper is engaging with the author in the way the author intended the work," I said.

"Oh, so you think that we should just let people write academic papers about *Gone With The Wind*, with 'Prissy the slave' and 'Big Sam's Paws,' do you think that that would be admissible material because 'the paper engages in the way the author intended the work', Jane-Marie?"

"It's not the same thing."

"Then please, tell me the difference. Explain to me why Joseph Conrad blowing up a person with mental challenges as an ironic climax is funny to you." Arden turned to the rest of the board. "When we permit literature with terrible values into our magazine, we can really hurt people. We can cause violence."

And so I lost. Piece #32 was scrapped. We bequeathed unto the righteous orator Arden whatever they wanted.

At the next meeting, Arden opened with, "I just want to address the fact that I no longer have a gender. You can use 'they' but I'm really struggling with this so just please bring as little attention to this problem as possible, alright? Just don't use pronouns in reference to me, but use 'they' if you have to. Does anyone have any questions? Does anyone have any problems with that?" Arden asked, staring at me.

In a writing class Arden commented on my friend's story, "I already told you how to fix your problems. If you don't fix the writing, then you're just wasting my time." Arden couldn't understand why my friend wouldn't want to change her writing based solely on Arden's preferences.

Arden was the favorite pupil of one of those holier-than-thou Robin DiAngelo professors. Arden was definitely a good writer, but Arden was not that much better than the rest of us. There are a lot of talented people in the world besides Arden. A lot of people have interesting stories to tell, they're just not as overbearing as Arden. It was not the content or the skill of Arden's writing, but the sanctimoniousness which made these two such a pair. In class, Dr. Tonya would look at a sea of raised hands. "Arden, what do you think?" She'd ask. "Arden, what's the problem here?"

No one wanted to go to the fundraiser. Arden said, "I actually can't come to the fundraiser because I have to work."

"Arden, we all have to go to the fundraiser," I demanded.

"I don't have my college paid for, so I need to work. What about you, Jane-Marie? Do you have your college paid for?"

One day I made a suggestion. "Can we have V-neck T-shirts for the club?"

"I need crew necks because they're gender neutral."

"Ah, so out of character for you to say something like that," I muttered under my breath.

"I'm sorry, what did you say?"

"Can I have a V-neck T-shirt, please? You can have a crew neck T-shirt." I snapped.

"Did you not hear what I just said?"

The president said, "No, we can have a combination," and I was relieved someone besides me finally wanted to talk back to this raging narcissistic lunatic. That night, I typed "Hey Arden, what about the T-shirts?" in our group chat. The next day I asked, "You put in a order for those T-shirts?" Over Spring Break I asked once more. "What about the T-shirts?"

Arden didn't respond.

When we ran for positions in the club the next semester, each board member had to give a little speech and step out so the other board members could adjudicate. Arden had to work most of those meetings, but Arden made sure to show up for mine. Arden claimed that I had discriminated against their gender, surprisingly. We had never had a one-on-one in-person conversation before, which means that their accusations were based solely on the V-neck crew-neck debate and other situations in the editing club that involved me defying them.

But Arden won.

No one stood up to Arden.

They voted me out of the editing club. They refused to let me know the exact accusations or let me respond. It occurred to no one, not even the adult advisors of the club, to let me speak in my own self-defense. Without ever asking me what happened, or letting me tell my side of the story, I was kicked out of performing a role for the organization I had dedicated so many hours of my life. Suddenly, I realized I was the bad guy. My friends cut me off. Alone, outside.

Six months after I got kicked off the board of the editing club, Arden courageously agreed to get coffee with me. By this time, they had cut their hair off and started wearing breast binders. They apologized for defaming and ostracizing me behind my back over t-shirts. Arden didn't take responsibility for their actions or put in any effort to make it right, but they did at least say sorry. So I wrote an open letter about this, because the advisers should not have allowed that to happen.

During COVID, my boyfriend lived in an apartment near Arden's, so I saw the transformation of their-to-his body. After Arden cut his hair short, his breasts disappeared. Arden didn't move home during the pandemic. I heard from a friend of the roommate that Arden's mom convinced the police to break into Arden's apartment. I heard that the mom broke in and had screamed at Arden. "Do you understand this isn't about you? I gave birth to a daughter! You were named after me!"

## EXHIBIT SIX:

During the George Floyd debacle, when everyone felt tremendous pressure to openly state their allegiance to Black Lives Matter on *Instagram* and *Facebook*, my roommate from my sorority at the time, Tsilia, wrote an angry message on Facebook. Tsilia's Jewish parents escaped the Soviet Union in the '90s and she felt like the social pressure was too similar. She wrote on *Facebook*, "I am sick and tired of these posts. You can't claim that every member of an entire profession is racist." Hundreds of students and teachers verbally abused Tsilia in the comments section. An *Instagram* page for Westmore racial justice ran a post saying that "Someone with [the exact scholarship Tsilia had] was a racist, and Westmore was showing their support for that racist by continuing her scholarship." Tsilia 'left' our sorority. Many of those who had once been her friend publicly denounced her. To be clear, Tsilia didn't say that Derek Chauvin did something okay. She said, "Not all members of the police profession are racist."

But everyone denounced, ostracized, and condemned her.

### EXHIBIT SEVEN:

Another girl named Reagan wrote something similar to Tsilia's post on her *Facebook* page. Reagan was treated in the exact same way. I contacted Reagan to video call. Reagan told me that in that semester she received the lowest grades she had ever gotten immediately after the post. Her professors saw her post and gave her C's in all of her final papers across the board—grades low enough to punish her but not too low that she could claim mistreatment, as she explained to me.

### EXHIBIT EIGHT:

My friend Addison from rural Alabama, who I met at Westmore, lost her mom when she was a kid. As the oldest sister of a big family, she adopted the role of second mom. Her dad turned to his community's church and engaged in redneck behaviors. Addison's dad freaked out when he saw her walking with her male friends from a church youth group.

Addison and I were in overlapping social circles. I observed that she exclusively got with men. Not gender-fluid or trans men. From what I observed, she only consorted with people who were born with penises.

Nonetheless, Addison from rural Alabama insisted that she was bisexual all of college. Need she sexually engage with women to be considered bisexual?

No, because she never did. What was she expressing then?

### EXHIBIT NINE:

On a corn maize wagon ride with my fiancé and his mom, four trans boys, born with XX chromosomes, sat behind us talking loudly. They were all white or white-passing. If I had to guess, I would say they were about four years younger than me, born between 2002-2004. Two of them were going bald, their hairline receding past the nape of the forehead.

One of them spoke louder than the others. When he chimed into their conversation, the rest of the wagon heard. A line I witnessed, my right hand to God:

"Well maybe he can be in our friend group then."

### EXHIBIT TEN:

I can count on two hands the number of people I know who claim to be something other than heterosexual who have never had sex.

### EXHIBIT ELEVEN:

One day I said that I knew a lot of bi people who have never had sex, and I was quickly told that this is a prejudiced observation. That merely saying that statement out loud constituted prejudice.

### EXHIBIT TWELVE:

In my middle school sex education, I saw 'abortion' at the bottom of the list of contraceptives, and assumed that it was a contraceptive. I knew that weird, boo-hick, crazy Christians didn't like abortion because they wanted to control women's bodies, but I didn't think about the issue much more than that. I accepted what I had been told.

In late college, I attended the Bodies Exhibit, which is a touring museum experience which shows the real muscles, skeletons, and nervous systems of human cadavers suspended in clear liquid. In the back of the exhibit, behind a curtain, it showed seven aborted baby girls, mostly from China, suspended in clear liquid at various stages of development. The littlest one was about the size of my thumb, but she had a clearly human cranial structure, although disproportionately large, with a little rib cage and arms and legs. At the end of the first trimester, the girl—because all the aborted babies on display were girls from Asia—was a little bigger than my stretched-out hand, and her head was developed with eyes, nose, mouth, fingers, toes, her intestines slightly spilling into the clear liquid through a puncture wound. In that moment, it dawned on me that in all those years that I was indoctrinated with the belief that abortion is another form of birth control, no one showed me a picture of a fetus.

After I exited the exhibit, I took out my phone, navigated to a search engine, and typed in 'aborted fetus.' No real picture came up. Drawings came up. Models came up. Nothing real came up. I was confused as to why no real images came up. Were they too gory? I typed 'whip marks' into the search engine and that black-and-white photo from the 1860s of Peter with keloid scars came up. Maybe whipping wasn't a good topic to evaluate the theory that the images of aborted fetuses had been scrubbed due to gore, so I thought of other gross things. I typed '3rd degree burns' and the search engine showed me what I wanted. I typed 'being stoned to death' and a video from Afghanistan of a woman hogtied in a hole getting bloodied up with rocks came up, for God's sake. The search engine hadn't scrubbed those search results! I typed 'aborted fetus' again and only diagrams came up. It seemed as though the search engine was hiding the images of the aborted fetuses.

Three years after I attended The Bodies Exhibit, I re-tried looking it up and real images came up for me this time. The search results had changed. Therefore, I think the more correct explanation is that: when I was in college, the search engine expected that I would want liberal search results, so it did not give me anything that might convince me to stop and think about the reality of what I believed.

You should take out your phone and try to find a picture of a real aborted fetus. Not spontaneously birthed early, not the aborted fetus of an animal, but a real aborted baby. If it's hard to find a photo like that, might that be because our gatekeepers in Silicon Valley have deprived you of information which would make you change your mind?

### EXHIBIT THIRTEEN:

In student work, I had a colleague named Emma who leaned on neuroticism with the rules. I liked her because she was honest, but she really went by the book and annoyed many of our coworkers. They started a GroupMe called "Fuck Emma" about my friend Emma. That is to say, a clique of twenty-two-year-olds did not have the moral development to refrain from participating in a group chat whose sole purpose was to bully and humiliate one of their colleagues.

### EXHIBIT FOURTEEN:

I wrote a story about Spruce Village foster care residential treatment. It's not a true story, obviously. But what is true is that five-sixths of the girls in the real home I worked at decided that they were not cisgender the same month.

### EXHIBIT FIFTEEN:

If, hypothetically, I met someone who came from money, who in childhood had been diagnosed with ADD or the like, who discovered in college that they were some gender expression that is not cisgender, and who claimed that they had been discriminated against, I would tell them that they have perverted class action politics. I say, "hypothetically" to be ironic. I probably have met more than one hundred people who fit that profile.

Before and during Martin Luther King Jr., a specific group of people with an immutable, genetic characteristic actually could not succeed in America's institutions. They needed class-action protection. Eighty years later, class action politics have been perverted to protect the children of the rich and connected and powerful.

A dirty little boy draped in furs stands in front of a hut. The sun is bright. His pupils shrink to tiny black dots in his pale blue eyes. His home is dug into the side of a hill, like a man-made cave. Unlike shelter built on flat land, a cave in the side of a hill prevents the possibility of attacks from behind and guarantees its inhabitants the high ground if thieves try to raid their home in the middle of the night. Close to the top of a slope means rainwater sloshes down and out, preventing flooding. The dirty little boy loved his home. His family had been so strategic when they built it.

The dirty little boy's uncle calls. "We've got long to go, now!"

His uncle substitutes for his father. Four summers ago, the dirty little boy's father had gotten into a fight over felling rights. The father's teeth were knocked out in that fight, so he had died of malnutrition that winter. The boy occasionally experienced a sudden flashback and saw his father gumming mushy apples like a fish. They embarrassed him—his memories.

"Did you hear me, brother's son?" the uncle calls to the boy. "Come on." The boy and his uncle are going to pay tribute to the Romans.

The boy had heard stories that Romans wear reflective metal shirts. They say that you can see your own death in their armor if you do battle with Romans. The boy tries to imagine his own reflection, and imagines that he looks like his uncle. He imagines that when he looks at a Roman, he will see his uncle's face in a metal shirt.

The boy and his uncle hike a two day's journey through rolling hills to the trading post, where their Reich collected taxes for tribute. The boy's uncle offers up a huge sack of rabbit and goat jerky to their leader. Three witnesses must be present to see that they have given their dues. To formalize the transaction, three intersecting lines are cut into a stick and given to the uncle as proof of receipt that he paid his taxes that year. Almost immediately after his uncle was handed his due stick, the hoofbeats of twelve riders pound in the distance. The boy steps out of the trading post and sees a glare of light, noticing his own reflection in their metal shirts.

The dirty little boy ignores what the Romans do, but instead focuses on his reflection. He realizes that he is gangly and round of face, with splotches of acne around his chin. Horrifically, he admits to himself that he resembles his father gumming apples much more than his strong, muscular uncle. On the periphery of his sight, the boy watches his Reich kowtow before the Roman soldiers, as the other men of his clan hand over their prized possessions and winter food for Roman taxes. Further still, as if from across a meadow, the boy hears a Roman soldier call his companion 'wirtuo.'

The boy mouths the funny-sounding word 'wirtuo' in his reflection on the soldier's shirt. "Wirtuo, wirtuo," he lips.

Finally the boy comes to the present, seeing his tribesman emasculated, and thinks, "This is crazy. Those little weird tan men in metal clothes should not get to take from us. They are milking us as if we are livestock. How can men so puny do that to my people?"

Five years later, as the dirty boy stands on the cusp of manhood, the Romans raise taxes so high that his tribe cannot pay. The Romans say that they will pay anyway and they come. They come over the hill which the boy has lived his whole life. They overtake the hill and ooze down it like an avalanche. The boy watches the army from inside his home, and it looks like standing behind a waterfall of metal shirts.

The dirty boy and his uncle could fight off three or four men with the highground, but not an army. Not a Roman army. His uncle dies fighting, sinews snapped, organs spilling out. The Romans tie the boy and his mother and siblings with a rope and march them south to a *lata-fungia*, where the dirty little Germanic boy works as a slave until his death.

Working on the plantation, the boy eventually learns that 'wirtuo' means manly, but something is lost in translation. In the boy's native language, 'manly' meant having strong arms and a thick neck, like his uncle and his father before his teeth were knocked out. To the Roman, though, wirtuo means being a good soldier. Wirtuo means having an attitude. The boy realized throughout his life in bondage that the Romans did not conquer with their strength, but with their strategy. Their masculinity rested in their minds. The dirty little boy, who is now a dirty Germanic slave in the boot of Italy, relives his capture a thousand times every day and obsesses over wirtuo, Roman manliness. Of course, it is written down as virtuo, but the Germanic slave does not know that because the Germanic slave cannot read.

The dirty Germanic slave pairs with a female—a Greek one, a beautiful woman with wide hips and curly hair. He impregnates this woman nine times, and four of the babies—all girls—live to adulthood. He tells his daughters to recognize wirtuo, and to choose men who conquer with their minds. He explains to his daughters that a man of wirtuo, is cunning, respectable, ruthless. A man of wirtuo is a winner. It was their wirtuo which caused him to suffer the indignity of being enslaved to tiny men, and don't make the same mistake, he warned his daughters. His people were only wirtuo in how they built their houses, like man-made caves inside hills. The Germanic slave in the boot of Italy dies of upset bowels at the age of forty-seven.

After many conquests and victories, marriages and rapes, after the Roman Empire implements Christianity and falls to chaos, that Germanic slave's 40th grandson would ride swordside of a king named Great Charles, Charles Magna, Charlemagne, the King of the Franks.

One day, Charlemagne sees a little girl, maybe seven years old, standing alone and ready to fight the coming army invading her village.

Charlemagne chuckles to himself, and says to his friend, "That girl has more wirtuo than any of you." In Charlemagne's evolved pronunciation of Latin, distorted by 400 years of illiteracy, he pronounces the word as 'virtuo' meaning 'possessing the spirit of a man.' The swordside rider is so impressed that he takes the brave girl as his wife in a child marriage, facilitated by a priest on the condition she bleed thrice before consummation.

The 25th grandson of Charlemagne's swordside rider would live in a English Monastery, speaking William the Conqueror's Norman French. The monk, with the same pale blue eyes as his Germanic slave ancestor, would see his brothers following the code of Saint Benedict obediently, and would call his companions 'virtuous,' meaning 'possessing good spirit.'

Sometime in the 15th century, the word virtually came to mean "possessing the same spirit as something else." As in *that boy is as virtuous as any knight I know.* That boy acts *virtuously,* with -ly stemming from the Germanic suffix -liech, which turns words into adverbs. Virtuous-ly. Latin, German, like the slave.

Armies fight as virtuously as their leaders.

You are virtually a leader.

You are virtually a Cretan, and should be put to death.

She is a child. No, she is virtually a woman. Have you seen the way she acts?

You are virtually the same as your brother.

The word wirtuo which Romans used to mean "possessing the spirit of a man" eventually became the English word 'virtually,' meaning "possessing the same spirit as something else."

Now, fifty generations after the word virtually came into the English language, the internet is invented. A fat boy hunched over a computer plays League of Legends, rotting his pale blue eyes in the radiation of screens. In his video game, the boy is in the mindset of war. He acts cunning, respectable, ruthless. He enacts the spirit of war, though not the reality of war. His masculine impulse to fight rests in his mind power.

Would his ancestor, the Germanic slave draped in furs, call his descendent wirtuo? And what of the word virtual?

The word *Taliban* means 'students' in Arabic. *Atloub* means 'I ask.' *Yatloub* means 'He asks.' *Talib* means 'asker.' And Taliban is the human plural. The US media never translated *Taliban*. It seriously would have tempered American morale for the US troops to square off against The Students.

How did The Students find themselves fighting a holy war against the United States? Well, starting around the '80s, child refugees from the Soviet invasion of Afghanistan gathered around to hear Wahhabi preachers claim that "God has kept you poor and ruined your government and punished you because you tolerate perversion. You tolerate indulgence. You keep pre-Islamic icons! You disregard God's justice! And for that, your family has been punished and you will be punished more."

"But who are we tolerating?"

"The perverts and the greedy!"

"But, I'm just a farmer's child. I have a mother and ten siblings. Who is it who has done so wrong?"

As the men listened to these speeches, the United States came in with their big metal machines and raped their land of its oil, soiling the fish in the rivers and harming their crops. The children became men and thought, "They say this is consensual trade, but I'm not making money off of it. They have ruined my ability to farm. Americans are gigantic. They're fat. They're greedy. They're sexually immoral. They're trigger-happy. They raped my land. And they mock my God. No, I do not accept world trade with the United States."

So, The Students saw their teachers, The Principles or, al-Qaeda, crash planes into the World Trade Center, and The Students cheered. Tons of people on this planet think that The Principles and The Students are the good guys. "The Americans came first. The Americans killed thousands of our innocent. Why should The Students not kill thousands of theirs?"

If we want to heal relationships in the world, if we want to stop war, we might start by translating their names.

Sometimes, I dream that I smell latex gloves and rubbing alcohol, but then I inhale fresh air and fall back asleep.

Every year until eighteen, I returned to the Hematology Clinic at Children's Hospital of Michigan, in midtown Detroit. This pilgrimage felt like visiting a museum. It certainly didn't feel like coming back to a place where I spent more than three years of my childhood.

After five years without a relapse, I was classified as a 'survivor.' I guess so. I am surviving.

My mom and I sat in the waiting room for four hours. There's only so long a person can stare at the floor. My leg oscillated, an involuntary response to impatience. I stopped, though, when I realized that the reason we were waiting so long was that I wasn't a patient of priority.

I rolled my eyes to the right and saw a toddler in a polka-dot dress sitting buried in her mother's bosom. The baby's head was lined with skimpy corn rows and dotted with patches of bare scalp. I watched her head as it moved with the rise and fall of her mother's breath, resembling the undulation of calm water. The girl pinched her mother's shirt and kneaded at the fabric like a cat, drifting into sleep.

The mother, a deeply maternal woman, made eye contact with me. She was strong and dressed in business attire. Our eyes met. She didn't drop her gaze, and although I should have smiled, I averted my eyes and grabbed a random pamphlet to read.

I scanned the pamphlet and turned to the second page, which read "Survivor Statistics" in bold letters. Although the pamphlet was about

a different kind of cancer than I had had, I started to read it anyway—
about how people have relatively low survival chances if diagnosed
before the age of one year and higher chances of survival if diagnosed
between two and seven years of age, and back to low chances for the
preteen and pubescent years. The statistics were represented by colored
silhouettes of people, like strings of paper dolls. To represent a fraction,
the last figurine had part of her face, her left shoulder and the flare of
her dress lopped off. Disturbed by this image, I put the pamphlet down
and returned to watching the room, taking careful note of the juxtaposi-
tion between movement and stillness.

The center of commotion was at the far back corner of the room—
the door that led to examination and treatment. A youthful nurse in pink
scrubs drew her legs to shoulder width and steadied herself before she
muscled the barricade open, escorting a family inside. Was it made of
solid steel? Each entrance and exit was accompanied by a soft *swish*-ing
sound—the kind of sound one releases during meditation.

"Ppshhhh," the door whispered as a boy around eight came through.
He walked topsy-turvy and had an ethereal gaze. He almost rammed into
another person, but his father gently steered him in the correct direction.
In his surreal mental state, he swayed back and forth. I knew exactly
what had happened to the boy. A spinal tap. Intrathecal chemotherapy.

I unconsciously started to mouth the words 'morphine' and 'versed';
two words I thought were one when I was a little girl, *morphine-and-versed*.
I smiled as I remembered jumping rope to the rhythm of that word.
Morphine is to relax you and make you disconnect from the fear and
pain. Versed makes you forget the agony of the procedure so you won't
dread it so much next time they have to do it.

A spinal tap, or lumbar puncture. One of the more painful things
that can happen to a young patient. You start by curling into a fetal position on your side. It took three people for me—two to hold me down—
because being immobile is imperative. If a child moves too much during
the procedure, the needle can damage the spine and potentially cause
paralysis. In this procedure, which I endured every other week for three

years, a nurse carefully inserted a hollow needle between the vertebrae in order to suck out a carefully measured quantity of spinal fluid for the sample. Then, she took an identical amount of medicine—chemotherapy fluid—and injected it through the same needle into the spinal column. They try to keep the volume of fluid in the spinal column constant. They say the patient is less likely to experience cripplingly painful headaches afterward if they keep the volume the same.

My mom told me that she felt ripples shoot through my muscles as the needle pushed in. If you've ever seen wild game shot in the head die, that's what spinal taps look like. Ripple seizures. My mom insisted, though, that she be the one holding me there—one of the greatest comforts she could give me.

Doctors must have spinal fluid to measure whether the cancer has relapsed in the brain. They insert chemo drugs into the spinal column to ensure it isn't blocked by the 'blood-brain barrier.' Even though I had more than seventy-five spinal taps during the course of my treatment, I don't remember the pain of the needle going into my back because versed is a benzodiazepine that erases the memory of pain. They don't give Versed to parents, though. My mom remembers everything. Some scars don't heal.

I envisioned his father holding him still and motionless. The boy was high, though, and ironically enjoying himself, spinning in happy and carefree circles. His father's body was hardened by physical labor and wrapped in a leather coat. The man seemed as tough as the door he just came through . . . except, that burly monstrosity was leaking tears. Maybe the lab report had revealed bad news. Or maybe he had simply watched his child suffer through the treatment one too many times.

I stood. My name had been called. *Good*, I thought. I couldn't bear staying still any longer. My fidgeting might momentarily cease. The fat nurse escorted my mom and me through the daunting door into the hallway for treatment. We were given a small room with a painted mural of a blue-haired mermaid to wait in. I sat in the room for three minutes until my antsiness consumed me and I fidgeted my way to the restroom because that was the only place I could go.

The bathroom reeked of regurgitated gastric acid. I could hear the girl hurling. The girl was most definitely a chemo patient: I could see the base of the IV standing in the gap between the stall door and the floor.

I listened, not disgusted. She had probably puked up her previous couple of meals and would probably regurgitate her lunches after. The smell of chemotherapy vomit is kind of like isopropyl alcohol.

I didn't want to go into one of the bathroom stalls. I just stood out of line, looking at myself in the mirror. Good, I was still standing there.

I looked at my hair. It was my best feature. Tight ringlets of shimmering gold. My aunt told me the universe gave me that color of hair because I won it. I took my victory, and my hair grew back the color of champions.

My eyes traveled down my body to my midriff. I observed my abdomen expand and contract, just like the chest of the mother in the waiting room. I inhaled deeply as the putrid smell wafted around me like an aura. I began to regret my escape to the bathroom as I started to feel sick myself.

Chemo patients have long-lasting aversions to food for years after the chemo is over. They used to make me eat a banana before I ingested the chemo, and my body associated the fruit with gut-wrenching nausea. More than six years after remission, the smell of the fruit would cause me to instantaneously vomit. Boom. In the fourth grade, an obnoxious girl at school opened a banana and chased me around with it. It was okay, though, because I puked all over her—ha!

My gaze in the mirror traveled down to my hips, which were strong and wide and built to bear babies, just like my mother's and her mother's before her. I was now staring at my legs. When I was in the third grade, I skinned my knee while playing four square. The abrasion leached blood down my leg and made small pools on the concrete. My reaction to the wound wasn't to scream, nor was it to run to the nurse. My reaction was to sit there and impassively examine my injury like an interesting artifact because having cancer teaches you to be separate from your body.

Really, the hair, the banana, and the complete break of mind-body connection were trivial complaints. Beauty comes from within. Some people don't like some foods. With a sovereign mind, I have time to nurture the academic and spiritual aspects of my being. Ironically, as a cancer survivor, I have more control over my own body than most.

If someone were to ask me to identify the most challenging aspect of leukemia, I would respond with the word 'isolation.' Leukemia is fundamentally a cancer of the white blood cells—the key cells in the immune system—so all the time one goes through treatment, the chemotherapy weakens and destroys the very cells meant to protect one's body from disease and infection. While the immune system is compromised, the victim can't go to public places that pose a high risk of spreading germs—places like schools and parks and birthday parties and ball pits inside arcades.

I'm an only child, so for the entire duration of my cancer, I stayed inside my insanely sterile house with either of my grandmothers. They organized a rotation between each other while my parents were working. They told me stories and played with figurines, which would embark on grand adventures fighting the world's injustices. My grandmothers were the kindest, most exciting companions one could have, but they weren't kids. And not having other kids in my life for years of my childhood irreparably changed the internal mechanism that makes me feel recognized by others.

Really, thank the Lord I'm an only child because if I had a sibling, that child would have had to be quarantined away from me or I might have had to spend three years of my childhood in a hospital room, with its awful smells and hourly check-ups by underpaid nurses.

Being separate from other kids made me introverted and, to a degree, intolerant of others. But, again, I've accepted it, because that time alone allowed me to feed my innate qualities without peer pressure. I learned to talk to myself, which led to a vast and deep internal sphere, which led to my writing, which led to this story—a story about a girl who continues to survive. Present tense.

I blinked and discovered that I had at some point started crying. The vomiting girl emerged from the stall. Was she ten? She cupped her hands under the faucet and rinsed out her mouth. The water in my ocular area started to recede. The girl noticed me. She stared into me with her bulbous, azure eyes. She looked away, looked back, looked away, and touched my hand.

I smiled softly, and the tears came back.

I grabbed her wrist, careful to avoid the intravenous needle that was stuck in her cephalic vein. I felt our energy unite. Us together. It was beautiful. The fluorescent lights of where we were faded into a pastel ether. It felt warm.

We just stared at each other's hands. Hers were white and so gaunt I could see every tendon. Mine were tan and vibrant and grown. She stared at my hands in absolute awe and wondered if hers would ever look like mine. She gently rubbed my knuckles and turned my hand over to examine my palm.

I knew she would survive because she was full of life. She was so radiant, she was channeling compassion and spirit and strength *into me*. She was projecting light like a star—like fusion was happening in her bones. She gently rubbed my hands and turned them over like she could read my palm.

With her wrist movement, her IV fell out, and her puncture wound started to discharge a stream of slightly off-colored blood. It wasn't an emergency; it was just a puncture wound. She jerked her hands away from me so she could apply pressure. She grabbed her IV, stood, and hurried out. I again felt alone.

I was standing in a hospital bathroom, my clothes permeated with the smell of rubbing alcohol and vomit, and all I could think of was how that beautiful girl wasn't standing next to me anymore. Two months after this checkup, I'd find an article online about that very girl.

May she rest in peace.

I was standing in a hospital bathroom, my clothes permeated with the smell of rubbing alcohol and vomit. I grabbed the sleeve of my shirt and started to knead the fabric like a cat. I watched my abdomen move in and out. I was still breathing.

I took a deep breath, past my diaphragm and into my solar plexus chakra. In and out. In and out. In and out.

I went back to the hospital room with the blue-haired mermaid. When Dr. Savason entered, he marveled at how grown I was, because he knew me personally, just like all of his patients. His Turkish accent proclaimed, "You're all straight," and he asked how my college experience was. Our meeting lasted less than seven minutes. I was done until next year. My mom and I got up to walk out.

The only thing standing between me and the rest of my life was that heavy metal door. I felt strong, almost warrior-like, as I pushed that barricade open and stepped outside.

I looked around the waiting room once again. The black woman and her baby were still sitting in the same seat as when I had left, but the girl was awake now. I made eye contact with her mother, and this time I smiled and walked over.

"Are you in remission?" the mother asked.

"I'm past five years in remission, so I'm cured."

The mother tucked her chin and kissed the toddler on her head. She whispered to her daughter. "You hear that, baby girl? That's what you'll be when you're her age, too."

*That's right*, I thought. *That's what she'll be.*

I felt whole and complete and confident. I left the hospital grateful to be alive and proud of who I was.

And very alert for bananas.

# SCREENS AND THE EGO: THE POEM

My Arab grandma believes in God
And she has a soul
I have mental emotional health

My grandma who believes in God laments
I'm depressed

My grandma who believes in God is seeking
I'm OCD

My grandma who believes in God needs redemption
I need better self-esteem

My grandma who believes in God bows to His purpose for her
I think it's stupid to think there's some dude in the sky

The therapist I see is trained in psychology
Psychology is the science of the soul
My grandma who believes in God
told me that science will never understand the human soul

My grandma who believes in God has a soul
I have screens and my ego